ESSENTIAL BIOLOGY

ESSENTIAL BIOLOGY

HERBERT T. HENDRICKSON

University of North Carolina, Greensboro

1817

HARPER & ROW, PUBLISHERS, New York

Cambridge, Hagerstown, Philadelphia, San Francisco,
London, Mexico City, São Paulo, Sydney

Sponsoring Editor: Steven Poe Heckel
Senior Project Editor: Rhonda Roth
Designer: Emily Harste
Senior Production Manager: Kewal K. Sharma
Compositor: Progressive Typographers
Printer and Binder: Halliday Lithograph Corporation
Art Studio: J & R Technical Services
Cover Photo: Emily Harste

Essential Biology

Library of Congress Cataloging in Publication Data

Hendrickson, Herbert T   1940—
   Essential biology.

   Includes index.
   1. Biology. I. Title. [DNLM: 1. Biology.
QH 308.2 H498e]
QH308.2.H46   574   80-24656
ISBN 0-06-042792-2

Contents

Preface

Every professional biologist that I have ever met is convinced that the most fascinating, profound, and relevant subject in the world is biology. As a confirmed member of this select group, I accept its major belief with no qualms. Our ability to earn a living while studying a subject that is both intellectually challenging and emotionally satisfying comes close to being idyllic. We relish the exquisite details of description and the precise vocabulary that allows us to communicate the almost infinite array of diversity that is found in the living world.

Biology is a very broad and diversified area of knowledge complete with a multitude of specific facts. Our contemporary understanding of the living world is based on, and relies heavily upon, insights gained from the areas of chemistry, physics, and mathematics. New information is being added to the subject at a stupefying rate. Most introductory courses in biology attempt to survey all of this knowledge within the span of one school year. Any way you look at it, that is a great deal of information to cover in a relatively short time.

For people who plan to major in biology or one of the allied areas, the standard introductory course is probably necessary. However, the majority of students taking an introductory course do not intend to become biologists. These people have great difficulty perceiving the significance and relevance of all the details that are usually presented.

One proposed solution to the problem of different needs of different students is to offer a second, shorter introductory course for the nonmajor. While many competent biologists may differ with this opinion, I feel that there are only two major functions that such a shortened introductory course must fulfill.

First, I think that this kind of course should portray clearly what science is and what constitutes good, scientific practice, how the practitioners of science work, and what they can and cannot do and why. An introductory course for nonmajors must spend more time dealing with the philosophy of science because these students may not be exposed to it again. I also think that it is a mistake to make vague, abstract generalizations about the way science or scientists work, without having a solid understanding of the factual material involved. Therefore, it is necessary to cite real examples of scientific work in enough detail to show how science differs from other areas of human knowledge.

Second, I think that this course should convey some knowledge from the body of information that is biology. Many students have come to the conclusion that biology, and probably the other sciences, involves nothing more than the memorization of a prodigious number of facts couched in a foreign vocabulary. It is easy to get that impression because it is, at least partially, correct. But biology is also a thinker's game. Biologists want to know the implications of those multitudes of facts and the relationships among them. That cannot be done without knowledge of the facts, but mastery of those individual facts does not constitute an understanding of biology. There is a tendency on the part of all of us involved in introductory courses to forget that.

These two basic ideas have led me to conclude that most introductory courses in biology for the nonmajor attempt to cover entirely too much information to do proper justice to any of it. Trying to survey the entire realm of biology results in frustration for everyone involved. Instead, I think the course should try restricting itself to a small number of general ideas or concepts, explaining them thoroughly enough so that the student can achieve real understanding of both the working of science and the principles of life.

This book is intended to be the text for such a course. It is designed specifically for a course offered at the University of North Carolina at Greensboro for nonscience majors. It differs from most biology texts in that it is smaller and less comprehensive. It attempts to promote ideas and discovery, rather than rote memorization.

While the use of specialized vocabulary has been reduced, it has not been eliminated, nor has it been simplified to the extent that it could be used by a fourth grader. This is supposed to be a college-level text. Similarly, it is not possible to discuss biology on a mature, intelligent level without some reference to chemistry, physics, and mathematics. These areas are only mentioned when they are pertinent to an increased understanding of biology.

This book will probably be considered easy when compared to many competitive texts on the market, but that does not necessarily mean that the reader will not be challenged to do some thinking. Some of the ideas that characterize thinking in the area of biology are not always intuitively obvious.

If you want to give an introduction to biology that will be manageable within the time span of a typical semester, it is obvious to me that something has got to be omitted. Because the principal orientation of this book is conceptual, the most notable exclusion is the traditional coverage of anatomy and physiology, or structures and functions. There is no discussion of muscular contraction, nervous impulse transmission, how blood flows through the heart, how food is digested, or other similar topics usually found in introductory biology texts.

The omissions do not imply unimportance. They are simply practical necessities. Studies of structure and function have generated a monstrous collection of names and details (all of which are very satisfying to professional biologists) but have resulted in very few conceptual breakthroughs. The major concepts that have resulted from such studies, such as negative feedback and the relationship between form and function, are explored in an appropriate context.

What I have tried to do in this text is to focus on those areas in which the most widely meaningful ideas have emerged. I hope that most students and their professors will agree with my choice of areas.

Herbert T. Hendrickson

ESSENTIAL BIOLOGY

1

The Scientific Study of Life

Science as a Discipline

Biology is the scientific study of life. One of the implications of that statement is that it is possible to study life in more than one way. However, within the covers of this book the multiple modes of investigation will be restricted to that which is scientific. There is an assumption that the contributions to human knowledge made by this scientific approach are in some way different from those made by other disciplines, such as history, fine arts, humanities, and mathematics. There is no claim made that these other approaches are better or worse, or more or less valid. However, they are different, and the insights and understandings of the world that we can achieve by using these different approaches will vary.

CHARACTERISTICS OF SCIENCE

The words *science* and *scientific* are familiar parts of the modern American vocabulary whose general meanings are vaguely understood but seldom defined precisely. This situation is partly due to the use of the term *science* in two different ways. First, science is a body of information about the material world; second, it is a process by which that information is obtained. Some people have described the process of science as nothing more than the rigorous application of common sense, and thus the content of science is that body of information obtained by using this rigorous, commonsense method. While there is a certain amount of circular reasoning in that description, there are few real conflicts in practice with respect to what is or is not science.

Science, as a body of information about the material world, deals only with those things we can observe either directly with our unaided senses or indirectly through some artificial aid to our senses. Furthermore, the observations must be repeatable and verifiable. A person does not have to be a believer or a practitioner of long standing to confirm the basic, factual observations of science content. It may sometimes require long training to operate some of the indirect aids used to make some observations, but the observations themselves are repeatable by anyone. This quality of consistency and repeatability in the making of observations is a characteristic feature of science. Unique events, or one-of-a-kind phenomena, are not the subject of science. Such subjects usually comprise the basis of history.

In addition, in science the observations are described in such a way as to minimize their ambiguity. Efforts are made to present observations as precisely and quantitatively as possible. For example, sizes of objects are usually measured by some mutually acceptable standard such as the metric system rather than subjectively described as large or small. Weights, volumes, velocities, temperatures, and many other physical properties are referred to by such standard-reference units.

The standard units of measurement used in scientific work are known as the metric system and often cause beginning students much difficulty. The difficulty is due to the fact that most of us have grown up with the English system of measurement, using inches, feet, pints, quarts, pounds, ounces, and several more obscure units. We are not used to working with any other system. This situation will probably change completely within one generation as the United States converts to the metric system. Legislation, enacted in 1975, encourages the country to adopt this alternative system of measurements within ten years.

Throughout most of the world the metric system is the standard means of measurement, because the metric system is simpler than our familiar English system. As an example, there are at least four commonly used units of length in the English system: the inch, the foot, the yard, and the mile, as well as several more or less obscure units such as the furlong, the rod, the chain, the fathom, and so on. The metric system has only one unit of length, the meter. The English system of volumes uses so many different units that you can easily be driven to distraction. For example, what are the relationships among teaspoon, tablespoon, cup, ounce, pint, quart, gallon, peck, bushel, barrel, and cubic yard? In contrast, the metric system has one unit of volume, the liter. The metric system has one unit of weight, the gram.

You would probably expect that one unit of length cannot be very useful for measuring something much smaller than the standard or something much larger. For example, we don't express the distance between New York and Los Angeles in inches, nor do we try to express the dimensions of a sheet of paper in miles. For this reason the metric system is set up in decimals. The standard units can be divided into tenths, hundredths, thousandths, and so on, or they can be expressed as multiples of ten. By multiplying or dividing the basic unit by tens, we obtain a continuous system of measurement from the exceedingly small to the immensely large. Each factor of ten that is used in conjunction with the standard unit can be represented by a standard prefix.

deci-	tenth	$\dfrac{1}{10}$	$\times 10^{-1}$
centi-	hundredth	$\dfrac{1}{100}$	$\times 10^{-2}$
milli-	thousandth	$\dfrac{1}{1,000}$	$\times 10^{-3}$

micro-	millionth	$\dfrac{1}{1,000,000}$	$\times 10^{-6}$
nano-	billionth	$\dfrac{1}{1,000,000,000}$	$\times 10^{-9}$
deca-	ten	10	$\times 10$
hecto-	hundred	100	$\times 10^{2}$
kilo-	thousand	1,000	$\times 10^{3}$
mega-	million	1,000,000	$\times 10^{6}$

From the listing we see that a thousandth of a unit ($\frac{1}{1000}$) is indicated by the prefix *milli-*. Thus a thousandth of a meter is called a millimeter; a thousandth of a liter is a milliliter; and a thousandth of a gram is a milligram. Ten millimeters ($10 \times \frac{1}{1000}$ meter) must equal a hundredth ($\frac{1}{100}$) of a meter. The standard prefix for a hundredth is *centi-*, so 10 millimeters must be the same as 1 centimeter. Conversely, 1 millimeter is the same as a tenth ($\frac{1}{10}$) of a centimeter. Long distances are usually expressed as so many thousand meters, or kilometers. Heavy weights are expressed as kilograms and large volumes as kiloliters.

The following conversions between the metric and English systems may prove helpful.

For length the standard unit is the meter.

1 kilometer (km)	3280.83 feet (ft) or 0.62 mile (mi)
1 meter (m)	39.37 inches (in.) or 3.28 feet or 1.093 yards (yd)
1 centimeter (cm)	0.3937 inch
1 millimeter (mm)	0.0393 inch
1 inch	2.54 centimeters or 25.4 millimeters
1 foot	30.48 centimeters or 0.304 meter
1 yard	91.44 centimeters or 0.914 meter
1 mile	1609 meters or 1.609 kilometers

For volume the standard unit is the liter.

1 liter (L)	2.11 pints (pt) or 1.06 quarts (qt) or 33.81 fluid ounces (oz)
1 milliliter (mL)	1 cubic centimeter (cm^3) or 0.03 fluid ounce
1 pint	0.473 liter or 473 milliliters
1 quart	0.946 liter or 946 milliliters
1 gallon (gal)	3.8 liters

For mass the standard unit is the gram (1 milliliter of water at 4°C weighs 1 gram).

1 kilogram (kg)	35 ounces or 2.204 pounds (lb)
1 gram (g)	0.035 ounces
1 ounce	28.4 grams
1 pound	453.6 grams or 0.45 kilogram

For temperature the standard unit is the degree Celsius (°C).

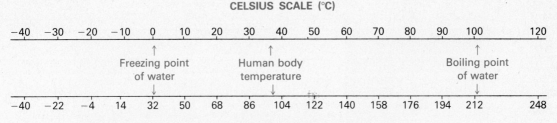

CELSIUS SCALE (°C)

| −40 | −30 | −20 | −10 | 0 | 10 | 20 | 30 | 40 | 50 | 60 | 70 | 80 | 90 | 100 | | 120 |

↑ Freezing point of water   ↑ Human body temperature   ↑ Boiling point of water

| −40 | −22 | −4 | 14 | 32 | 50 | 68 | 86 | 104 | 122 | 140 | 158 | 176 | 194 | 212 | | 248 |

FAHRENHEIT SCALE (°F)

The simplicity and elegance of the metric system cannot be approached by the English system. No longer is it necessary to remember all the different conversion factors needed to go from one unit to another. In the metric system everything is expressed in one unit to some power of ten.

Items of study, either objects or processes, must be referred to precisely and unambiguously so that others will be able to repeat the observations. Thus an extensive vocabulary has evolved in science, which beginning students frequently find overwhelming. The purpose of all the new words is not to confuse or distract; the purpose is to enable people to communicate clearly and specifically.

If science consisted of nothing but the accumulation of specific verifiable observations, it would indeed be one of the least interesting of human endeavors. A recitation of particular facts is relatively dull. Human beings, in general, have a strong tendency to seek order and coherence from the chaotic world of particulars and to try to establish relationships among the isolated facts. When patterns of observations appear, people often construct generalizations. For example, you could easily go out, make a series of observations, and report that every healthy adult chicken, turkey, parakeet, canary, duck, and pigeon that you saw was covered with feathers. Those observations are individual, repeatable, and verifiable. From these observations you might infer that all healthy adult chickens, turkeys, parakeets, and so forth, are covered by feathers. You might even be inclined to go further and say that since all six of these kinds of things are called birds, then all healthy adult birds are covered with feathers (see Figure 1.1). Formulating a generalization of this sort is a perfectly normal, human activity, and it is also good basic science.

With the construction of a generalization a good scientist will insist upon testing its validity. In general this means making further observations. The best generalizations are those that are phrased in such a way that their authenticity can be easily subjected to scrutiny. If in the test of the generalization, observations are made that contradict the

Figure 1.1

(a)

(b)

(c)

Each of the healthy adult organisms pictured has a body covering of feathers. Each of these organisms is known as a kind of bird. You could generalize that healthy adult birds have a body covering of feathers. (a) Gannets; (b) flamingo; (c) Laughing Gulls; (d) Toco Toucan; (e) Mallard; (f) Barn Owl. [Photos (a) and (b) courtesy of Dr. John B. Hess. Photo (c) courtesy of John E. Wiley. Photos (d)–(f) courtesy of Matthew E. Stockard.]

(d)

(e)

(f)

generalization, then it cannot be accepted as valid. The generalization must be rejected or modified in such a way as to take into account the discrepant observations. If, on the other hand, repeated testing reveals no contradictory observations, our confidence in the validity of the generalization is increased. Obviously the more often that observations are made that are consistent with the generalization, the more confident everyone will be that the generalization is correct.

The degree of confidence in the validity of a generalization is often expressed by the use of different words. For example, a **hypothesis** is a relatively new, insufficiently tested generalization for which there is only weak confidence in its correctness. It is a generalization or a conclusion that is consistent with some factual observations, but not enough observations to make anyone confident that it is most probably correct. A generalization that has been tested repeatedly and found to be consistent with all factual observations to that point encourages a much higher degree of confidence in its validity. Such a generalization is often called a **theory.** When the confidence of a generalization's validity reaches the level at which the observations are of such consistency and regularity that they can be expressed as a mathematical formula, the generalization is often called a **law.**

It is worth noting that there is nothing rigid about the distinctions among hypothesis, theory, and law as described in the previous paragraph. These words indicate varying degrees to which the scientific workers who know the basic facts think the statement is correct. The more significant distinction is probably between a hypothesis and a theory. A hypothesis is certainly consistent with some observations, but it is relatively untested and not generally accepted. On the other hand, a theory is a hypothesis so well substantiated that it is generally accepted as correct.

There are two points that are important to stress at this time. The first is that scientific laws (or theories, or hypotheses) are descriptive generalizations of the material world as it exists and as it most probably will continue to exist. They are summary statements of observable reality, which say nothing about how the world should be or how we think it ought to be. There are no statements of moral judgment made in science. The content of science may be used to make moral judgments, but this activity transcends the realm of science. The second point is that it is not really possible to prove a scientific generalization. One can amass thousands of observations consistent with a generalization, which leads to great confidence in its validity, but that is not proof of its absolute correctness. On the other hand, a single contradictory observation is enough to disprove a scientific generalization.

One of the results of this tentative descriptive nature of science is that its practitioners have a tendency to be very cautious when making

Table 1.1
Frequency of Birth in Weeks After Last Menstrual Period

WEEK	FREQUENCY	WEEK	FREQUENCY
28	0.0016	38	0.0909
29	0.0016	39	0.2000
30	0.0019	40	0.2857
31	0.0041	41	0.1764
32	0.0041	42	0.0833
33	0.0074	43	0.0294
34	0.0086	44	0.0135
35	0.0172	45	0.0071
36	0.0256	46 and more	0.0071
37	0.0454		

generalized statements. This caution expresses itself in several different ways. One way is to very carefully proscribe the conditions under which a particular generalization is known to be valid. Another technique is to avoid the use of dogmatic terms such as *always* and *never*. Biologists, in particular, are characterized by the use of such cautious words as *usually, often, most of the time, there is a tendency, frequently,* and so forth. A third example of this caution is in the use of probabilistic statements. Many natural phenomena seem to defy absolute generalizations and yet are consistent enough to be somewhat generalizable. Predictions of weather appear to fall into this category, as do many biological events. For example, you can generalize that a pregnant woman will give birth to a baby (as opposed to her giving birth to a puppy or a kitten). You can also generalize that she will most probably give birth within 35 to 44 weeks after her last menstrual period (data in Table 1.1). You can even generalize to the point of saying that half the time a pregnant woman will give birth to a son and half the time to a daughter. Each of these is a valid generalization that can be made, but each carries a different level of certainty based on probabilities.

What all this means is that much of science, including biology, requires a knowledge of some observable facts, because all generalizations (hypotheses, theories, and laws) are dependent upon those observable facts, not vice versa. Keep this basic point in mind as you work your way through some of the major biological concepts. The generalizations are meaningless without the facts to support them.

TESTING GENERALIZATIONS

Having established that the rigorous testing of a generalization is a crucial function in differentiating science from other approaches to knowledge, we now must determine how hypotheses are tested. There are two basic ways in which hypotheses might be tested: repeated observation and experimentation. Either one of these approaches can effectively discredit the validity of a given generalization. In a very real sense the best way to gain confidence in the validity of a particular generalization is to look for evidence that contradicts it. In other words, a good test of a hypothesis should be designed to find its shortcomings. Consequently, scientists must be open and honest in their search and evaluation of all facts that are relevant to a given generalization. They (and we) are not allowed to ignore any data.

Repeated Observation   The technique of repeated observation is so simple and basic an approach that it is often overlooked as a fundamental tool of science. Multiple observations of the same phenomenon are much more likely to be accepted as valid than are single observations. In the same way, multiple observations bearing on the same generalization increase the level of confidence in the validity of that generalization. For example, take the generalization, made earlier, based on six observations that healthy adult birds have feathers. Our confidence in the validity of this generalization would be increased substantially by observing more examples of healthy adult birds to see if they too had feathers. The greater the number of examples of birds with feathers, the greater will be our confidence in the validity of the generalization. If, however, we examined a healthy adult bird and found it totally lacking in feathers, we would have to reconsider the generalization and either reject it or modify it in some way.

Experimentation   Experimentation is usually considered a more advanced form of hypothesis testing in that it is a system of controlled observations. An experiment is supposed to limit the number of variables involved in an observation to one. Any differences in outcome can be assigned to the difference in the variable. The idea of comparison is central to the nature of experimentation. The whole purpose of an experiment is to assign a cause to an effect when different observations are made under different circumstances. By minimizing the number of factors that may be responsible for causing variable observations, the experiment maximizes the chances of correctly identifying the factor that produces the changed results.

Let me clarify that abstract explanation with a concrete example. Suppose you have a hypothesis, based on some previous observations or prior knowledge, that handling a live toad will cause warts to grow on the hand. An experiment can be designed that will challenge the validity of this generalization. Take a large number of people and divide them into two groups. It is important to divide the people without any deliberate aim or purpose so that each group is not identifiable by some characteristic feature. Each of the two groups should have essentially the same composition of people with respect to factors such as age, sex, racial origin, general health, level of education, religious beliefs, and so forth. Each member of one of these groups, which will be called the **control group,** must handle something other than a toad and be specifically prevented from handling toads during the experimental testing period. The members of the second group, called the **experimental group,** must handle a toad. At the end of some appropriate period of time all the members of both groups should be examined for the appearance of new warts. A comparison of the number of people in each of the two groups who acquired warts during this period of time should reveal whether or not there is a relationship with handling toads. Unless there are many more people in the experimental group who have acquired warts than there are in the control group, the hypothesis has been contradicted and must be either abandoned or modified (see Figure 1.2).

There are some fine points of discrimination in this particular experiment that are worth elaboration. The hypothesis claims that toad handling will cause warts. Technically any observation of a person who has handled a toad without getting warts constitutes a contradictory observation that invalidates the hypothesis. However, this obser-

Figure 1.2

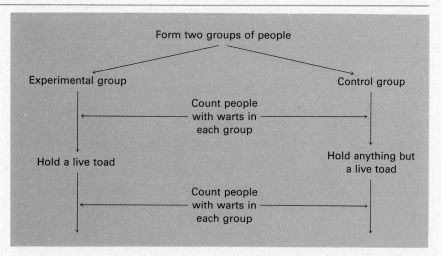

A simple diagram of how to design an experiment. Hypothesis: If you hold a live toad, you will get warts. The hypothesis is supported only if many more people in the experimental group get warts, compared to the number of people in the control group who get warts.

vation still allows for the possibility of the alternative hypothesis that some of the time handling a toad will cause warts. Furthermore, the hypothesis does not claim that toad handling is the only cause of warts. Therefore, the observation that someone who has not handled toads has acquired warts does not constitute proof that the hypothesis is incorrect. This possibility of other causes for warts requires the use of a large enough sampling of people in both groups to allow the expression of non-toad-induced warts. The control group is needed to obtain an accurate measure of this phenomenon. The observation that some people in the experimental group got warts does not constitute support for the hypothesis unless their numbers are significantly greater than those in the control group whose warts were induced by some other (unspecified) cause.

The critical words in the last statement are *a significantly greater number*. When are two numbers significantly different? Presumably, if one were to sample a population many times to measure some property, such as the incidence of warts, it would be unreasonable to expect to obtain precisely the same number or proportion in every sample. Some variability in values would be expected due to pure chance. The more often such a population is sampled, the more confident one becomes of the limits to this variability due to chance. The areas of mathematics concerned with probability and statistics have developed methods for estimating these limits. When two values obtained by sampling differ by more than one would expect within the limits of chance variation, they are said to be significantly different. Explaining the mathematical intricacies of statistics is not within the realm of this text. However, such probabilistic reasoning is becoming more and more common in most areas of science, and hence an understanding of statistics can be extremely valuable as a tool in interpreting scientific data.

To summarize, science is a body of information about the material world, which has been amassed by rigorous, critical observational methods and has been subjected to continuous testing and retesting.

The Concept of Life

Biology is a subdivision of science that deals with the phenomenon of life, which raises yet another interesting question. What is life? Because of the various restrictions that have been imposed by the previous description of science, answers such as "Life is a bowl of cherries" or "Life is a pink pomegranate floating in a sea of prune yoghurt" are not adequate. They may be valid and express deep, metaphysical insights, but they do not reflect verifiable, observable reality. Furthermore, the statements do not appear to be readily testable.

Everyone knows what is meant by life (see Figure 1.3). All students have experience with the idea and the reality of life through the fact of being alive themselves. Despite the great familiarity of the idea, it is almost impossible to define the word in any noncircular way. Invariably people end up resorting to the use of words such as *living, alive, not dead,* or *animate.* Most of you probably learned a long time ago in some English class that it is not acceptable to define a word by using that same word or a derivative of it. Thus it is extremely difficult to define what separates living from nonliving.

I have chosen to make the distinction between *living* and *nonliving* intentionally to avoid the narrower alternatives of *living* and *dead.* Dead objects are certainly nonliving, but the reverse is not necessarily true. In its most common usage the word *dead* implies that the object in question was at some prior time alive and that it has now ceased to be alive. The term *nonliving* does not carry this connotation and hence is a broader, more inclusive term. The very real problems associated with determining whether an individual human being, for example, is alive or dead are only a part of the general difficulty in defining life.

Typically what is done when definition is not possible is to characterize those traits and properties shown by things that are generally considered living and compare them with nonliving entities. In other words, if it is generally possible to recognize things that are alive, what traits (other than being alive) do they all share that nonliving things do not share?

Composition   One of the simplest (and in retrospect perhaps simple-minded) approaches is to determine what living things are made of and compare this to the composition of nonliving things. This experiment has been done many times and it has been discovered that living things are composed mostly (about 98 percent) of the **elements** carbon, hydrogen, oxygen, and nitrogen. The remaining 2 percent or so of living things consist of the elements calcium, phosphorus, sulfur, iron, zinc, sodium, chlorine, fluorine, copper, potassium, iodine, and just about all the other elements known to occur naturally on this planet. There is nothing materially unique in the elemental makeup of living things that differentiates them from nonliving things. They are all composed of the same elements. There are no new special elements found only in living things. However, there is a difference in the proportions of the elements found in living things compared with nonliving things. Comparison of the elemental composition of the entire earth with that of liv-

Figure 1.3

Which of the items illustrated are considered living and why? (a) Galapagos Tortoise; (b) Kermit the frog; (c) maple tree; (d) rock; (e) automobile; (f)lichen; (g) Opossum (not playing). [Photo (a) by the author. Photo (b) © Henson Associates, Inc., 1956, 1981. Photos (c)–(e) and (g) courtesy of John E. Wiley. Photo (f) courtesy of Dr. John B. Hess.]

Table 1.2
Elemental Composition of Living and Nonliving Units (in percentages)

ELEMENT	HUMAN BODY	SEAWATER	EARTH'S CRUST	ATMOSPHERE
Oxygen	25.5	33.	47.	21.
Hydrogen	63.	66.	0.22	0.01
Carbon	9.5	—	0.19	trace
Nitrogen	1.4	—	—	78.
Calcium	0.31	0.01	3.5	—
Phosphorus	0.22	—	—	—
Chlorine	0.03	0.33	—	—
Potassium	0.06	0.01	2.5	—
Sulfur	0.05	0.02	—	—
Sodium	0.03	0.28	2.5	—
Magnesium	0.01	0.03	2.2	—
Silicon	trace	trace	28.	—
Aluminum	—	trace	7.9	—
Iron	trace	trace	4.5	—
Titanium	—	trace	0.46	—

ing things reveals that the planet has a disproportionately greater supply of some elements such as silicon, iron, and aluminum than is found in living things (see Table 1.2).

These observations imply two things. First, living things are not just a grab-bag sampling of the building materials available on this world. There is (or has been) something selective involved in determining how much of which elements become incorporated into living material. The second implication, based on a little knowledge of chemistry, is that it seems reasonable to assume that the elements in living things are combined together differently from the way they are combined in the nonliving world. Pursuing this latter idea leads us to some new and interesting observations.

Individual units of elements (called **atoms**) combine to form structures called **molecules.** Molecules may be composed of two or more atoms of the same element, in which case they are called elemental molecules. Nitrogen gas (molecules composed of two atoms of nitrogen) and ozone (molecules composed of three atoms of oxygen) are examples of such elemental molecules. Molecules may also be composed of two or more atoms of different elements, in which case they are called **compounds.** Water is an example of a compound; it is composed of two atoms of hydrogen and one atom of oxygen, often represented as H_2O.

An examination of the composition of living things at the molecular level reveals not only many elemental molecules, and compounds

Table 1.3
Some Examples of Organic Compounds

CARBOHYDRATES	Glucose	Glycogen
	Galactose	Ribose
	Lactose	Maltose
	Cellulose	Starch
	Fructose	Chitin
	Sucrose	
LIPIDS	Fats	Waxes
	Steroids	Cholesterol
	Oils	
PROTEINS	Muscle	Cartilage
	Hemoglobin	Albumin
	Insulin	Gelatin
	Enzymes	Horn
	Hair	Skin
	Fingernails	Eye lens
NUCLEIC ACIDS	Deoxyribonucleic acid	
	Ribonucleic acid	

that occur commonly in the nonliving world, but a tremendous array of compounds that occur naturally only in living things or in things that at one time were alive. The vast majority of these compounds contain the element carbon, and because of their restricted occurrence in living material, they are called **organic compounds.** Organic compounds by and large can be subdivided into four major categories called **carbohydrates, proteins, lipids,** and **nucleic acids.** Some examples of these kinds of organic compounds are listed in Table 1.3. As a general rule, then, living things are composed of organic compounds while nonliving things are not.

Knowing something about the chemical composition of an object can help us to differentiate living from nonliving, but we are still left with a very incomplete comprehension of the concept of life. Compositional analysis may show good, solid differences between a live mouse and a marble statue of a mouse. On the other hand, it may not reveal significant differences between a live mouse and a statue made of the right combination of organic compounds. Furthermore, comparing the compositions of a live mouse and a dead mouse will give the same results, thus making no distinction at all between living and nonliving. To make matters worse, when we perform the compositional analysis, the live mouse ends up dead. It should be apparent that compositional analysis may be helpful, but it is probably not the most useful technique for distinguishing living from nonliving.

Organization   It would appear, then, that there is something special not only in the way that elements are put together into molecules to make living things but also in the way that the molecules are arranged and maintained together. Anything that disrupts this organization of all the constituent parts causes the termination of the phenomenon we call life. It also seems that the way the parts are organized imposes some restrictions on the size that living things can achieve. Every living organism ever found has been within the size range of a half of a micrometer to a hundred meters. It would seem that the arrangement of molecules precludes making a living thing outside this size range. In addition, this organization of constituents appears to impart several functional properties that characterize living things.

Functional Properties   Living things maintain a property called **metabolism.** Metabolism is the sum of many chemical **reactions,** some of which are building complex molecules and some of which are breaking down complex molecules. Usually the reactions that result in large molecules (called synthetic reactions or **anabolic** reactions) require the input of **energy.** Those reactions that result in smaller molecules (**catabolic** reactions) usually release energy. Metabolism is the sum of anabolism and catabolism. The energy released by catabolic reactions can be used to drive anabolic reactions, but the total amount of energy released by catabolism is often insufficient unless either more large molecules are taken in from the outside world or an external energy supply can be tapped. Thus living things can be described as users of an external energy source, which take in material and incorporate it into their structures while simultaneously giving off material to the external world. Another way of saying the same thing is to say that living organisms are systems in a state of dynamic equilibrium with the external environment. The idea of dynamic equilibrium is important because it says two things at once. Not only do living things appear to remain relatively constant and distinct from their immediate surroundings, but they do so while constantly exchanging materials with that environment.

Living things characteristically increase in size (within limits) through some span of time, and this phenomenon is called **growth.** The growth of living things is almost always from within the total organism rather than caused by the simple addition of more material to its periphery. In this respect the growth of living things is fundamentally different from the growth of crystals or geologic structures such as **stalactites** or **stalagmites.**

All living things are potentially capable of producing new individuals that appear very much like themselves and are, in turn, capable of performing all the functions carried out by the predecessor. In other words, living things are capable of **reproduction.** They can make little ones of the same kind. Sometimes it requires the cooperation of two individuals to complete this function; sometimes it doesn't. Once again that sounds like a neat, exclusive definition of life, until you recognize that it is an adequate description of the shattering of rocks by exposure to subfreezing temperatures. Obviously the breaking up of rocks is something else (although the past fad of keeping pet rocks does give one pause for thought).

Living things also have the ability to react to various kinds of **stimulations** from the external world (something pet rocks cannot do). Stimulation usually means some kind of external change, although it can include internal changes due to the effects of metabolism. Sometimes the reactions are obvious, as when you step on a cat's tail or hit a cow with a cattle prod. Other times they are much more subtle, as when you pick a flower or change the direction from which sunlight shines on a plant. This ability to react to changes in the surroundings is usually called irritability.

Finally, the characteristics possessed by the different kinds of living things have a strong tendency to change over long periods of time. This acquisition of new properties, coupled with the loss of old properties, is called **adaptation** or **evolution** (the two terms are almost completely interchangeable). This ability of living things to be modified through time is largely due to the combination of slight imperfections in the systems of reproduction and differential reproduction and survival.

If you put all these characteristics (with all their elaborations, clarifications, and exceptions) together, you still probably wouldn't have a very good definition of *life*. Perhaps you would have a better description of life, and you could get a start on achieving a better understanding of what living things are, but you would not have a very good definition. Two of the goals that I hope to achieve in this book are to give you a fuller appreciation of life (regardless of how, or even if we manage to, define it) and to point out some of the subtle interactions and complexities that are necessary to maintain this extraordinary concept that we call *life*.

Summary

1 / Science is the ordered knowledge of the material world obtained through verifiable and repeated observations that have been subjected to repeated testing.

2 / *The metric system is a decimal system of weights and measures using the meter as the fundamental unit of length and area, the liter (one cubic decimeter) as the fundamental unit of volume, and the gram as the fundamental unit of mass or weight.*

3 / *Scientists have developed an extensive vocabulary in order to maximize specificity and minimize ambiguity in communication.*

4 / *Scientific generalizations are usually called hypotheses, theories, or laws. Hypotheses are generalizations consistent with observations but relatively untested. Theories are hypotheses that have been repeatedly tested and found consistent. Laws are hypotheses that are so consistent that they can be expressed mathematically.*

5 / *The methods of science cannot really prove anything. All scientific methods can do is demonstrate consistency between factual observations and explanations. Many generalizations and explanations must be probabilistic.*

6 / *Hypotheses may be tested either by repeated observations or by experimentations. Experimentations are comparative observations that differ by only one factor.*

7 / *Life is not a concept that can be easily defined in unambiguous terms.*

8 / *Living materials are composed largely of the elements carbon, hydrogen, oxygen, and nitrogen. These elements are organized into organic compounds called carbohydrates, lipids, proteins, and nucleic acids.*

9 / *The organic compounds of living materials are organized in such a way that the functional properties of metabolism, growth, reproduction, response to stimuli, and adaption are normal outcomes.*

Questions to Think About

1 / Would people be more inclined to float through the air if the law of gravity were only a theory of gravity? Why or why not?

2 / Suppose the U.S. Weather Service predicts a 90 percent probability of rain with temperatures around 5°C for the day that you've planned a picnic and a swimming trip. Uncle Smedley says it has never rained or been cold on this day for the last forty years. What do you do? What does this decision-making situation have to do with science?

3 / Advertisements on television, radio, magazines, and so forth are great places to find generalizations that are not usually substantiated by scientific facts. Brand X deodorant will make everyone like you; Brand Y milk is healthier for you; Brand Z product will complete whatever is missing in your life. Design an experiment that will test the validity of the claims in your (least?) favorite commercial.

4 / Can you construct a concise definition of the word *life* without using some derivative of the word?

5 / A flame is not generally considered to be alive, yet it seems capable of performing most of the functions that we associate with living things. Why isn't a flame considered alive?

6 / Trees and people are considered living, but rocks are not. What characteristics do trees and people share in common that are not possessed by rocks?

2

Cellular Basis of Life

Introduction

In the first chapter, where I discussed the characteristics of life, I mentioned something about the organization of constituent parts. What I said at that time was not a very complete explanation of how living things are organized. It was sufficient to demonstrate that there is something special about the way living things are organized, but the nature of that organization was not particularly important at that time. Because the nature of the organization of living things is unusual, and because it is one of the major discoveries in biology, I think it should be explained more completely. The purpose of this chapter is to clear up some of the mysteries concerning the structure and organization of living things.

To one degree or another, all of us are familiar with the kinds of organisms that we can readily perceive, and there is every reason to believe that this has generally been true in the past. In effect what I'm saying is that the general public's familiarity with life forms today is on approximately the same level of technological sophistication that it has been at any time in the past. There is a multitude of **organisms** that we see with great frequency, such as dogs, trees, grasses, and insects. We have a rough idea of what these living things look like, and therefore we establish a vague conception of what living things are supposed to be like. This working model has probably been modified by the less frequent encounters with more unusual kinds of creatures, such as the things you see on a trip to the beach, the zoo, or a tropical greenhouse. Other living things make their presence known through senses other than sight, such as the call of a Whip-poor-will or a katydid, or the scent of skunks. Most of us also have experience with organisms whose existence is known because when they are present we feel and are ill. In addition to all these more or less readily perceivable kinds of organisms, there are many organisms humans are not equipped to perceive directly or easily.

MICROSCOPY

One of the limitations of the human visual apparatus is the size an object must be before it can be perceived as an object. There is some individual variability in this regard, but for the most part if an object is smaller than a tenth of a millimeter, it is not visible. A related problem is that of resolving power, or **resolution.** If the distance between two points is less than a tenth of a millimeter, they appear as one object. A familiar problem posed by the limitations of resolution is encountered by people who drive at night on long stretches of straight roads. The headlamps of approaching vehicles invariably appear as one point of light initially, and it is impossible to resolve whether a one-lamped motorcycle or a two-lamped automobile is approaching until the distance to the approaching vehicle has been shortened considerably (see also Figure 2.1).

The fifteenth-century discovery that curved pieces of glass—or, more properly, lenses—could be used to make objects appear larger,

Figure 2.1

The inset of this photograph has been enlarged greatly to reveal that the image is composed of many closely spaced dots. The unaided human eye is unable to resolve the small open spaces between some of the dots, so they appear to be continuous. [Photo courtesy of Matthew E. Stockard.]

increased the variety of objects that were visible to people and also improved resolving power. To a great number of people lenses constituted little more than an interesting toy that provided some harmless amusement. There were also practical effects, like the invention of eyeglasses. Several people, such as the Italian Galileo Galilei, arranged lenses together in series, which made distant objects appear closer and larger and therefore more susceptible to study; that is, they invented the **telescope.** Thus Galileo was able to get a much better idea of what the moon looked like, and his sponsor, the duke of Medici, was able to get a better idea of what his enemies were doing outside the city walls.

In a somewhat similar fashion, several mid-seventeenth-century

Figure 2.2

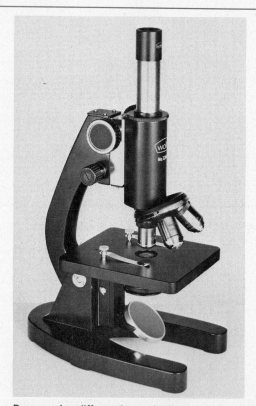

By arranging different lenses in the proper sequence at precise distances, we can make very small items appear much larger. Such an instrument, called a microscope, has contributed immensely to the study of living materials. A modern microscope of the sort used by millions of biology students is illustrated. [Photo courtesy of Carolina Biological Supply Company.]

scientists invented the **microscope,** in which an appropriate serial arrangement of certain kinds of lenses could create a very effective kind of magnifying glass and enabled the easy perception of even extremely tiny objects. A modern example of such an instrument is illustrated in Figure 2.2. Given a very primitive extension of the visual senses, Antoni van Leeuwenhoek, a rich Dutchman of some leisure, proceeded to examine everything imaginable through his lenses. Of particular interest to van Leeuwenhoek was the examination of many kinds of fluids, such as blood, pond water, semen, and sweat, because within these he saw many kinds of unusual little things, moving around as if they were alive. These things were invisible to the naked eye, but with the microscope there existed another whole world populated by what he called "tiny animalcules" (see Figure 2.3). In standard scientific fashion van Leeuwenhoek then proceeded to publicly announce his findings of all these wonderful little creatures. His discoveries prompted more people to try duplicating his observations, and they also were rewarded by finding more unusual-looking creatures.

Figure 2.3

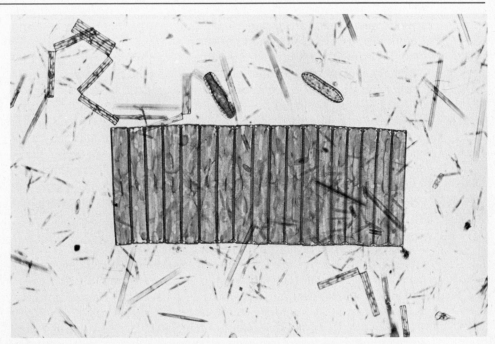

Most samples of pond water will often reveal an abundance of microscopic organisms. Here we see a mixture of organisms called diatoms, which are not visible to the naked eye. [Photo courtesy of Carolina Biological Supply Company.]

DISCOVERY OF CELLS

Once the word got out, many people started looking through microscopes and describing what they found as well as giving names to the different sorts of creatures. One of these people was an Englishman named Robert Hooke. In 1665, for some reason that has become obscure, Hooke tried looking through his microscope at a cork. Cork is simply the bark of a particular kind of oak tree, and it possesses a number of properties that make it ideal for sealing narrow-mouthed bottles. If you have ever tried looking at a cork under a microscope, you know that all you can see is a big black nothing. Hooke used an exceptionally sharp penknife to carefully slice off some very thin slivers of cork and he examined these. He found that the cork was made up of millions of tiny boxlike structures, which resembled monk's cubicles or prison cells. Therefore, he called them **cells.** This was the first time the word *cell* was used to refer to a unit of structure of living material (see Figure 2.4).

THE CELL THEORY

As time went on more people made microscopes, looked at things through them, noticed new things, and told their friends and colleagues about them. More cells were found in other kinds of plants and in animals, too, but not in rocks, metals, or other nonliving things. By 1838 a German **botanist** named Matthias Jakob Schleiden decided that he had looked at enough material to announce publicly that, as far as he could tell, all plants were made up of these little boxlike structures called cells. In the next year a German **zoologist** named Theodor Schwann, who had been working quite independently of Schleiden, concluded on the basis of his own studies that all animal material is composed of cells. If one assumes that all living things are either plants or animals, then the conclusion is unavoidable that all living things are composed of cells. This conclusion has become known as the cell theory, and Schleiden and Schwann are given joint credit for its original formulation.

Those of you who are skeptics ought to be squirming around trying to verbalize an objection to this simplistic generalization. It would seem that those little creatures found by van Leeuwenhoek are neither plant nor animal, yet they did seem to be alive. Admittedly, there was and still is a great deal of confusion over what to call those organisms that can only be seen under a microscope, but regardless of what they are called (we'll discuss that later in Chapter 7), they are alive and they are composed of cells. The chief difference between van Leeuwenhoek's "animalcules" and the sorts of organisms we generally see and deal with is the number of cells involved, or, more precisely, the total size of the organism. Microscopic organisms consist of a small number

Figure 2.4

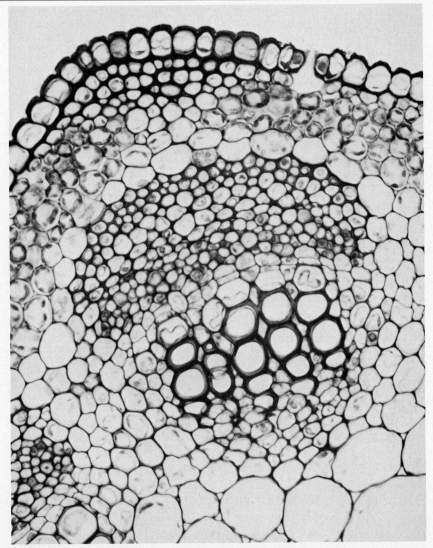

An alfalfa stem has been sliced across and magnified 810 times to show the basic cellular structure of living material. Notice that all the little boxlike structures are not of the same shape and size. [Photo courtesy of Dr. John B. Hess.]

of cells, frequently as few as one, which carry out all the characteristic functions of living systems. At the other extreme, elephants and oak trees are composed of millions of individual cells packed together in a cooperative fashion, resulting in one living organism. In effect, enough evidence had been gathered to formulate the hypothesis that cells are the basic, functional or structural unit of all living things.

In my opinion this generalization qualifies as an excellent hypothesis, not because of its correctness but because it is so easy to conceive of ways to test it. Carrying out the tests may be time-consuming, but the basic procedure is straightforward and simple. All that has to be done is to find an organism that everyone agrees is alive and show that it doesn't have a cell. If that can be done, then the hypothesis is demonstratively false and must be abandoned.

So far, every living thing that has been looked at, which includes every living thing that is known, has been found to be composed of one or more cells. The testing of this hypothesis is so obvious, and has withstood testing so many times, that there is now a high degree of confidence in its correctness, and consequently it is called the cell theory.

The cell theory had achieved such a high level of acceptability by 1858 that another German, Rudolf Virchow, expanded it to say something more about the basic nature of living things. Virchow hypothesized that cells are not only the fundamental unit of living things but are produced by previously existing cells. In short, Virchow proposed that the only way a living cell could be brought into existence was for an already existing cell to reproduce itself.

SPONTANEOUS GENERATION

Virchow's expansion of the cell theory challenged a controversial topic that had been simmering for at least four hundred years. What the adherents of the cell theory maintained was that only living organisms that are composed of cells can produce more living organisms, which are also composed of cells. This statement denied the possibility of producing any kind of cellular organism spontaneously out of noncellular or nonliving materials, for which there was a long history of supporting folklore.

Most of the sorts of organisms reputed to be the result of this process of **spontaneous generation** were of the disreputable kind (by human standards), such as toads, worms, maggots, and mildew. These are usually organisms associated with or responsible for decomposition, spoilage, putrefaction, and so forth. Recipes existed in old texts that enabled people to produce their very own loathsome creatures. If, for example, you take one sweaty horse blanket and 5 pounds of barley to the back corner of a dark shed, sprinkle them with bat blood on a night with a full moon, mumble the right magic words three times while a dog howls, then wait one week; you will produce your own rats. Alternatively, you can leave some meat out on a counter and it will generate its own population of maggots. Moist leather will generate mildew and molds. Broth will bring into being its own spoilage organisms.

Most people today will chuckle over such an explanation and label it as superstitious nonsense, but we are still faced with the difficulty of proving that spontaneous generation does not occur. Actually, many experiments had been performed prior to the nineteenth century that weakened the idea of spontaneous generation as a valid explanation. All the experiments followed the same basic format of excluding the organisms in question, as can be demonstrated with the following example.

Place equal amounts of raw meat into two widemouthed jars. Cover the top of one jar with a piece of cheesecloth, and place both jars outside where they can be exposed to sunlight and fresh air. At the end of a week or two the jar without the cheesecloth will probably have maggots crawling all over the meat, while the jar covered with cheesecloth will be maggot-free. This experiment was performed in the mid seventeenth century by an Italian named Francisco Redi. Redi noticed during the experimental period that flies were attracted to both samples of meat. While the flies were free to crawl on, eat, and, as Redi also observed, deposit eggs on the meat in the control jar, they were excluded from the experimental meat by the cheesecloth. Redi concluded that the maggots resulted from the contact of flies with meat, and he subsequently showed that maggots are the larval offspring of the eggs laid by the flies in the meat. Fortunately, an occasional frustrated fly would deposit its eggs on the cheesecloth, and these would subsequently hatch out maggots that were also excluded from the meat by the cheesecloth (see Figure 2.5).

Figure 2.5

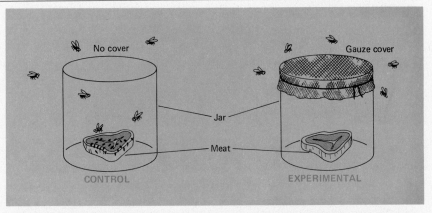

Redi's experiment. If flies are excluded from the meat, no maggots are produced.

Similar kinds of experiments performed with different sorts of organisms and other basic ingredients demonstrated repeatedly that spontaneous generation did not occur. In all cases the resulting organisms could be shown to have been produced not by the special nature of the ingredients but by previously existing organisms making use of the ingredients.

By the middle of the nineteenth century, spontaneous generation had been generally discredited as a causal explanation for the existence of all organisms except **microorganisms,** those visible only through a microscope. Here there seemed to be some doubt. It was generally known that you could destroy the microorganisms in a nutrient broth by boiling, and if the container were then sealed, it would remain uncontaminated. However, if the seal was broken, the broth would quickly spoil due to the action of living microorganisms. Adherents to the spontaneous generation hypothesis maintained that this was due to the generation of life by the interaction of air and nutrient broth.

Figure 2.6

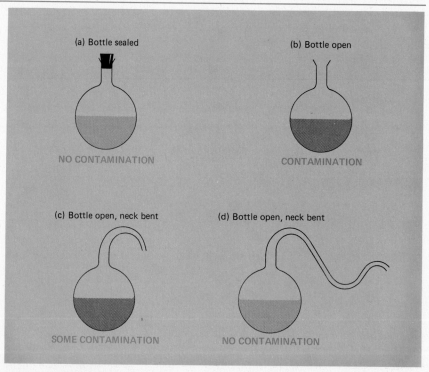

Pasteur's experiment.

It took an ingenious set of experiments by the Frenchman Louis Pasteur to finally put spontaneous generation to rest. Pasteur reasoned that there must be microorganisms in the air that are responsible for the spoilage of the broth. The problem is to devise a way that allows the broth to be in contact with air and yet still excludes the microorganisms. His solution was to heat the necks of the broth flasks so that they could be bent and drawn out into long tubes (see Figure 2.6). If the necks of the flasks were bent over so that the opening was directed downward [see Figure 2.6(c)], microbial contamination and spoilage still occurred but at a much slower rate than if the neck remained upright. Apparently, while many of the organisms settled out of the air directly, some did drift through the air in directions other than downward. By bending the neck into S curves, and simultaneously decreasing the diameter of the neck [as shown in Figure 2.6(d)], Pasteur allowed the passage of air to the broth but impeded the movement of microorganisms so effectively that the broth did not spoil.

Having thus disposed of the idea of spontaneous generation, we can say, with a very high level of assurance, that all living things are composed of cells, and all living cells are produced by previously existing living cells.

Cellular Organization

Just what is a cell? Cells are very difficult to describe in general because there are thousands of different kinds and they all look and act differently. Trying to describe a generalized cell of which there are no actual examples invariably results in either including some feature not found in all cells or excluding some features possessed by many kinds of cells. In either case you wind up with an impression that is probably incorrect. But in spite of the difficulties we still need some description of a cell. As a very imprecise, general analogy, cells are usually like tiny cellophane bags filled with a viscous glop. The most familiar, and probably least typical, example of a cell is the yellow part in the middle of a hen's egg. Several other examples are shown in Figure 2.7.

Now that you have a mental picture of a cell, it's time to formalize the vocabulary. The cellophane bag that holds everything together is called a cell **membrane** or a plasma membrane. The glop inside is a mixture of many different things in a watery medium collectively referred to as **cytoplasm** or protoplasm. You may also see it called cell sap. In other words, a cell consists of a plasma membrane surrounding cytoplasm.

Figure 2.7

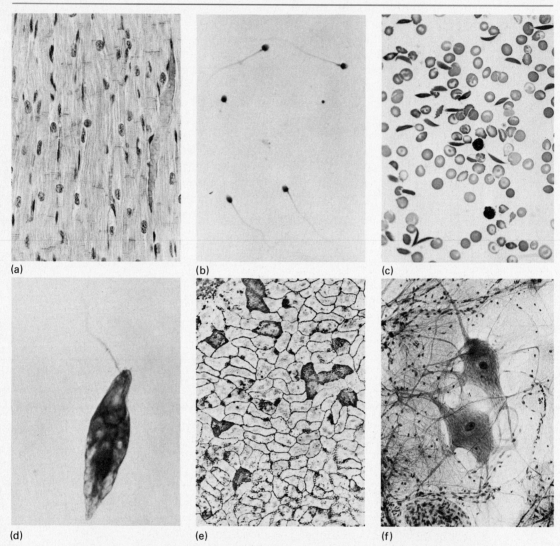

(a)   (b)   (c)

(d)   (e)   (f)

Describing the typical cell is a nearly impossible task because of diversity. Notice in particular the many differences exhibited in the shapes and sizes of these eucaryotic cells. (a) Heart muscle cells; (b) human sperm cells; (c) red blood cells; (d) *Euglena,* a single-celled organism; (e) mesothelial cells; (f) spinal nerve cells. [Photos courtesy of Carolina Biological Supply Company.]

MEMBRANES

Membranes form the boundary between what is outside the cell and what is inside; thus the cell's membrane is the point of physical contact with the surrounding environment. Its integrity is necessary to maintain life. As I indicated earlier, living cells are in a state of dynamic equilibrium with their environment, taking in some materials and giving off others. These materials must pass through the cell membrane. It has been found that the cell membrane is a very selective barrier; it allows passage of some materials while restricting or even prohibiting the passage of others, and the specific properties may vary depending upon the direction of flow or transport of the substances. For this reason it is called a semipermeable membrane. Some of these properties of permeability are almost certainly due to the physical arrangement of the molecules of which the membrane is composed, forming something

Figure 2.8

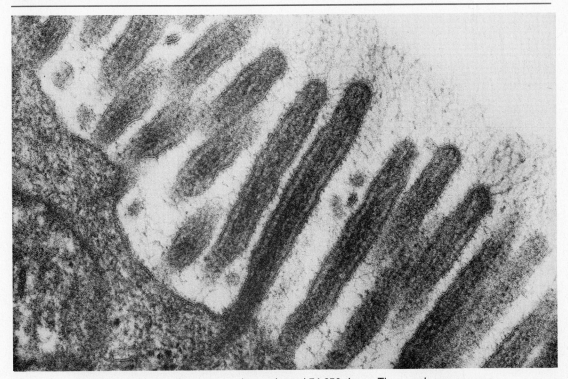

Plasma membrane of a mouse colon, enlarged 74,650 times. The membrane appears as a double line separating the dark area of cellular contents from the light areas of space. [Photo courtesy of Dr. D. W. Misch.]

like a molecular sieve. Other characteristics of permeability appear to be the result of specific chemical reactions associated with the membrane, reactions that use energy to force substances from one side of the membrane to the other. A very high resolution photograph of a plasma membrane is found in Figure 2.8.

CYTOPLASM

The fluid inside a cell, the cytoplasm, is mostly water with many chemical substances in either **solution** or **colloidal suspension** (see Figure 2.9). It is a very rich mixture, composed of hundreds of different kinds of molecules. Most of the smaller molecules are said to be in solution. A solution is a uniform mixture of two substances that cannot be separated by ordinary physical means. The components of a solution may be either solid, liquid, or gas. Their individual molecular sizes are so small that they are capable of randomly dispersing with respect to one another, forming a stable and unique set of properties. Common examples of solutions include seawater, the atmosphere, soft drinks, and metal alloys.

Figure 2.9

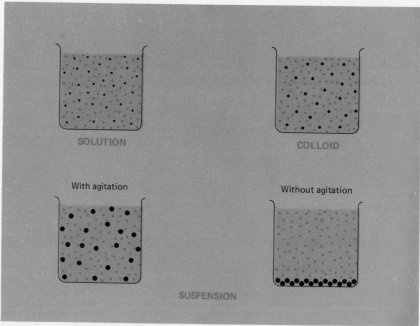

SOLUTION   COLLOID

With agitation   Without agitation

SUSPENSION

Kinds of mixtures.

At the opposite extreme of mixtures is the situation called **suspension.** A suspension is a nonuniform mixture of two or more substances that can only be maintained by agitation, which is a form of energy input. The difference in size of the two components in a suspension almost always results in their separating into discrete parts. Familiar examples of suspensions include muddy water, vinegar and oil, and dusty air.

Between these two extreme forms of mixtures there is a nebulous intermediate called a **colloid.** Colloids are nonuniform mixtures in which the components are not distinct enough to separate. Because one of the components tends to be much larger than the other, colloids are usually more viscous than are solutions, and they are also not as stable as solutions. In many respects colloids appear to be permanent, nonagitated suspensions. Depending upon physical conditions, colloids may be fluid or semisolid, and some kinds of colloids are able to change back and forth from one phase to the other. Homogenized milk, gelatin, and egg white are all familiar examples of colloids.

The colloids and solutions of which cytoplasm is composed are usually in motion, streaming around within the confines of the cell membrane. This cytoplasmic streaming can be seen particularly well in many kinds of algae and fungi. In certain **amoeboid** organisms it constitutes the basis of a form of locomotion (see Figure 2.10). The huge number and variety of kinds of molecules found in cytoplasm frequently react with each other to produce new substances. These new substances either form new colloids, new solutions, or insoluble cellular components, or are eliminated from the cell through its membrane. This array of chemical reactions within the cell constitutes a description at the molecular level of the processes of metabolism, growth, reproduction, and irritability, which are the characteristic features of life.

PROCARYOTES AND EUCARYOTES

Before saying anything more about what can be found in the cytoplasm of cells, we must make a fundamental distinction between two different kinds of cells. There are many, mostly microscopic, organisms whose cellular structure consists of little more than a cell membrane surrounding a nondescript cytoplasm. These organisms may also have various kinds of embellishments on the outside of the membrane, such as wall-like coatings and whiplike projections, but no distinguishable structures floating around in the cytoplasm. These organisms are said to have **procaryotic** cells and are called procaryotes or monerans. Among the living examples of procaryotic organisms are all the bacteria and a group of greenish, photosynthetic creatures that are variously known as blue-green algae or blue-green bacteria.

Figure 2.10

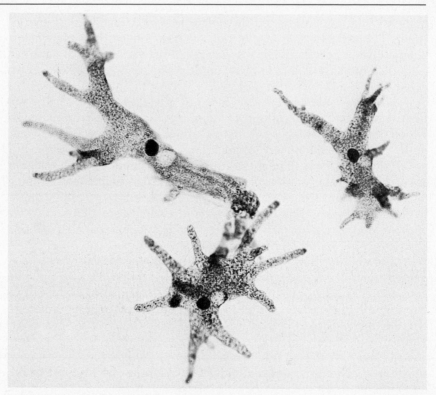

The three little blobs are unicellular organisms called *Amoeba proteus.* The cytoplasm in each cell pushes against the cell membrane, producing footlike extensions. The ebb and flow of these cellular protrusions enable the *Amoeba* to move and engulf food materials. [Photo courtesy of Carolina Biological Supply Company.]

The second kind of cell is called the **eucaryotic** cell, and all living organisms except bacteria and their blue-green relatives are composed of one or more eucaryotic cells. Eucaryotes can be distinguished easily by the presence of membrane-bound structures called **organelles** within the cytoplasm of the cell. There are many different kinds of organelles, which have recognizably distinct structures and seem to perform separate activities within the cell. Usually it is not possible to find all known organelles in any one cell, but almost all eucaryotic cells will have a **nucleus,** and many will have one or more other organelles. Some of the major differences between procaryotic and eucaryotic cells are shown in Figure 2.11.

Figure 2.11

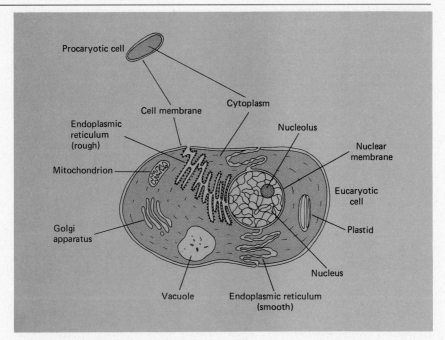

Comparison of procaryotic and eucaryotic cells.

NUCLEUS

The nucleus of a cell is a relatively large, spherical organelle, separated from the surrounding cytoplasm by a double-layered membrane called the **nuclear membrane.** The nuclear material within the nuclear membrane appears to be denser than cytoplasm and reacts with a **Feulgen stain** to produce a distinctive red color. The staining properties of the nucleus are caused by a high concentration of a nucleic acid called deoxyribonucleic acid, or **DNA** for short. Because DNA, in conjunction with the stain, is responsible for the production of a bright color, the material within the nucleus is often referred to as **chromatin.**

On certain occasions the chromatin appears to condense into small sausagelike bodies called **chromosomes.** At the time the chromosomes appear, the nuclear membrane is not apparent, so the nucleus does not exist as a discrete entity. These events correlate with the replication of the cell. Consequently, it is thought that one of the functions of the nucleus is the control of cellular replication. I will have more to say about this subject in Chapter 4.

Figure 2.12

These three individual organisms are called *Acetabularia.* Each individual consists of a single cell. [Photo courtesy of John and Linda Curtis.]

There is also a reasonable body of evidence indicating that the nucleus is crucial in controlling the normal functioning of a cell. Presumably the nucleus directs and regulates cellular activity. **Enucleated** cells (those from which the nucleus has been removed) do not function normally. Such cells continue to perform some activities properly for a limited period of time, but eventually they stop performing those functions or carry out others at inappropriate times. What results is a progressive deterioration to cellular death. The only way that an enucleated cell can be restored to its normal functioning is by the reintroduction of a normal nucleus, which is an extremely tricky operation that must be done before the degeneration is too advanced. Enucleated cells have never been found to replicate themselves.

The best evidence supporting the idea that the nucleus controls the functioning of the cell comes from a series of nuclear transplant experiments that have been performed on a large single-celled alga called *Acetabularia*. *Acetabularia* is green, lives in salt water, and is shaped like a toadstool or an umbrella about 4 centimeters tall (see Figure 2.12). We can recognize and talk about three different regions of the cell: the base, the stalk, and the cap. The nucleus is usually located in the lower part of the stalk. If the upper part of the stalk or the cap is carefully removed, the base and lower stalk will regenerate a new top to the cell that looks just like the original. If the nucleus is also removed, regeneration will not occur. Of the several different kinds of *Acetabularia* that are known, two may be distinguished by the shapes of their caps. Both caps are round, but one is smooth while the other is fringed.

What would happen if you removed the caps from both kinds of *Acetabularia*? From what I have told you so far, you should be able to guess that the stalk that used to have a smooth cap will grow a new smooth cap and the stalk that used to have a fringed cap will grow a new fringed cap. From this we can conclude that bases of *Acetabularia* cells will grow their own distinctive form of caps. An interesting difference is observed when we not only remove the caps but also transplant nuclei into stalks that have had their own nuclei removed. A smooth-capped stalk with a new smooth-capped nucleus still regenerates a smooth cap, but a smooth-capped stalk with a fringed-capped nucleus now generates a fringed cap. A fringed-capped stalk with a new fringed-capped nucleus continues to regenerate a fringed cap, but a fringed-capped stalk with a smooth-capped nucleus generates a new smooth cap (see Figure 2.13). We are forced to conclude that it is neither the membrane nor most of the cytoplasm in the base that determines the shape of the cap but something special within the nucleus that organizes and controls the cytoplasm and membrane to conform to some plan. In a very real sense the nucleus can be thought of as a blueprint, operating plan, and control center for the entire cell.

Figure 2.13

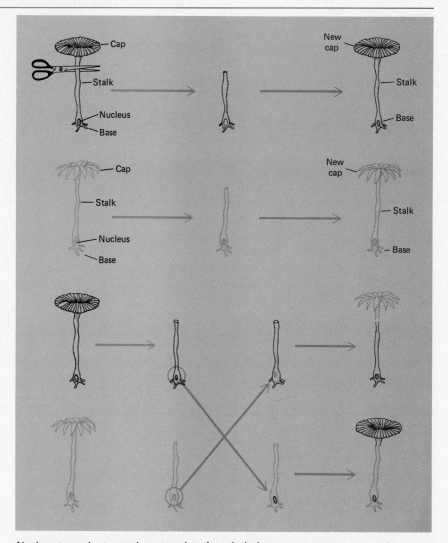

Nuclear transplant experiments using *Acetabularia.*

OTHER ORGANELLES

There are many other inclusions of the cytoplasm that perform various specific functions and are called organelles. It requires several volumes the size of this text to describe completely the present knowledge of organelles and new contributions to this knowledge are being made

Figure 2.14

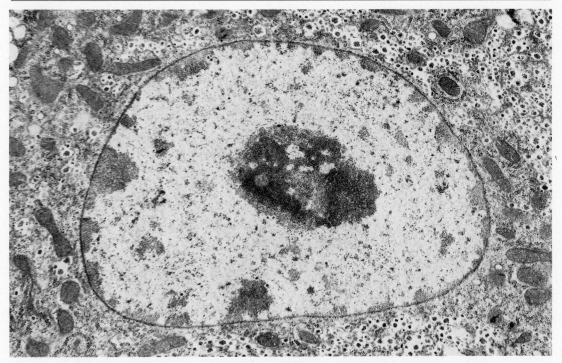

Nucleus of a mouse liver cell, enlarged 12,000 times. The large dark area near the center of the nucleus is called a nucleolus. [Photo courtesy of Dr. D. W. Misch.]

nearly every day. However, a brief description of some of the major organelles should be sufficient to provide an idea of their diversity and some appreciation for the internal complexity of eucaryotic cells.

Endoplasmic Reticulum   The **endoplasmic reticulum,** often called the ER, is a series of convoluted membranous tubules that appear to form a tortuous pathway between the exterior of the cell and the nucleus. It is, in fact, an extension of both the cell membrane and the outermost of the two membranes surrounding the nucleus (see Figure 2.14). Frequently, small dense bodies called **ribosomes** are found associated with the ER, giving it a rough, granular appearance (see Figure 2.15). This complex of ER and ribosomes has been shown to be the site where new protein molecules are assembled. Ribosomes can be found also in procaryotic cells, often associated with the cellular membrane, where they are also involved with protein synthesis.

Figure 2.15

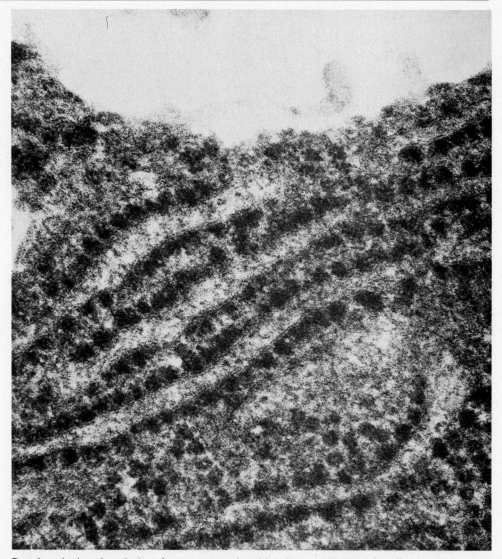

Rough endoplasmic reticulum from a mouse thyroid, enlarged 166,900 times. The intense black spots are ribosomes, which are attached to the outsides of the membranous tubes, which are the ER. [Photo courtesy of Dr. D. W. Misch.]

Mitochondria   A **mitochondrion** is a small organelle surrounded by two plasma membranes, the innermost of which is folded inward (see Figure 2.16). Each cell usually has one or more mitochondria, which

Figure 2.16

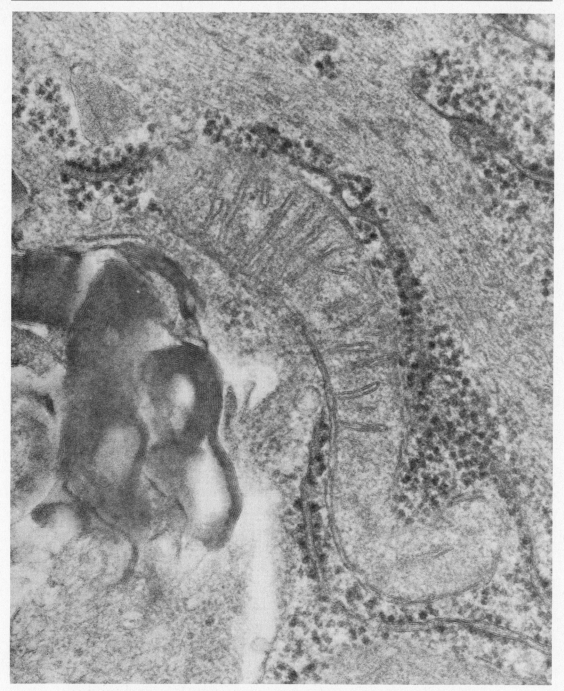

Mitochondrion from a mouse thyroid, enlarged 90,000 times. Note how the inner membrane of the mitochondrion is folded inward, forming partitions within the organelle. [Photo courtesy of Dr. D. W. Misch.]

drift around in the cytoplasm. Usually the more active a cell is, the more mitochondria it has. Mitochondria function as energy converters within a cell and are often referred to as the powerhouses of the cell. Mitochondria are packed full of **enzymes** associated with the infolded membrane, and these enzymes accelerate a series of reactions called

Figure 2.17

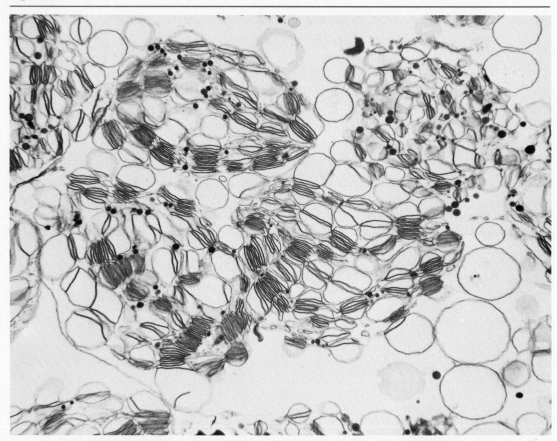

Isolated chloroplasts enlarged 12,000 times. The dark areas of the organelles, which are stacked like sandwiches, are where photosynthesis takes place. Many of the clearer areas within the organelles are starch grains, which have been produced by photosynthesis. [Photo courtesy of Dr. D. W. Misch.]

aerobic respiration. More will be said about this in Chapter 9. Mitochondria also carry a small unique supply of DNA.

Plastids There are several recognizably different kinds of **plastids,** which perform different functions. All of them are small double-membraned organelles found only in plants and the plantlike algae (see Figure 2.17). Some seem to serve the function of storage of insoluble materials such as starch, while others contain special pigments used to trap energy. Further consideration of these pigments and their involvement in photosynthesis will be found in Chapter 9. Some plastids, like mitochondria, also carry their own small supply of a unique DNA.

Vacuoles **Vacuoles** look like membrane-bound bubbles within the cytoplasm. They come in assorted shapes and sizes and perform several different functions. They are used commonly to store many kinds of substances and as chambers within which food substances are digested [see Figure 2.18(a)]. Because the enzymes responsible for food digestion can also digest many of the major constituents of the cytoplasm, it is advantageous to keep this process in the relative isolation of vacuoles. There is also a class of vacuoles in some kinds of organisms that regulates the amount of water in the cell by pumping out any excess [see Figure 2.18(b)].

Golgi Apparatus The **Golgi apparatus,** or the Golgi bodies, is a collection of highly convoluted, greatly compacted membranous tubules (see Figure 2.19). For a long time it was difficult to determine what, if anything, they do. It is now thought that they are responsible for the production or packaging of many different kinds of **secretions.** Secretions are substances, released by a cell, that perform some extracellular function. The process of secretion is not to be confused with the process of **excretion,** which is the removal of metabolic waste products.

Multicellular Organisms

It is important to keep in mind that because cells are the fundamental unit of all living things, it is perfectly reasonable to expect that a single cell can perform all the functions that we think of as being characteristic of life. There are, indeed, thousands of kinds of living organisms that consist of nothing more than single cells. Because an organism is a living thing as it occurs in nature, it is possible for a cell to be an organism and for an organism to be a cell. As I've already mentioned, cells

Figure 2.18

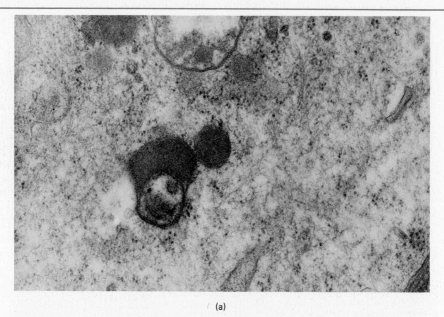

(a)

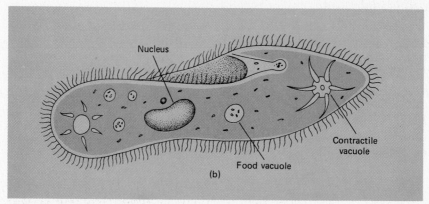

(b)

(a) Lysosomes from a mouse colon, enlarged 100,000 times. Lysosomes are specialized types of vacuoles. [Photo courtesy of Dr. D. W. Misch.] (b) Diagram of the single-
celled organism *Paramecium,* showing a contractile vacuole.

tend to be too small to be readily seen by humans and recognized as
organisms. The implication of this last observation is that most of the
things that we do recognize as organisms must consist of many cells.

Multicellular organisms are phenomena with which we are all
familiar. Even if this familiarity is on a relatively superficial level, it is

Figure 2.19

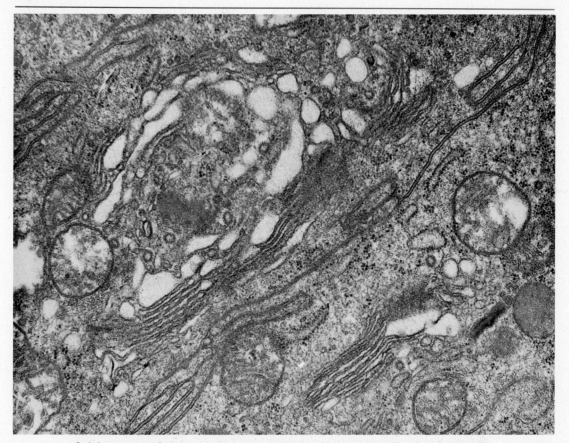

Golgi apparatus from a mouse colon, enlarged 120,000 times. The Golgi apparatus looks like an isolated, smooth endoplasmic reticulum that has been severely compacted. [Photo courtesy of Dr. D. W. Misch.]

enough to make us aware of two important generalizations regarding multicellularity. The first generalization is that there are a tremendous number of different kinds of multicellular organisms. The second generalization is that the multicellular organisms appear to play a much more prominent and dominant role in the workings of the living world; they seem to be much more important in the general scheme of things than are the single-celled organisms.

One obvious interpretation of these observations is that multicellular organisms are bigger and easier to see by humans and thus appear to be more diverse and more important, but only from our point of view.

Perhaps if we could more readily see single-celled organisms, we would find much more diversity in and discover a much greater degree of importance played by these creatures. While recognition of this potential source of bias is valuable, it turns out to be inadequate. Biologists have been trying to tally up the number of different kinds of organisms for over two hundred years, and while the count is still not very precise, it is clear that the number of kinds of unicellular organisms is exceeded by the number of kinds of multicellular organisms by a factor of at least 15 (i.e., there are 15 times as many kinds of multicellular organisms as there are kinds of unicellular organisms). Furthermore, in terms of producing effects on this planet that are noticeable, of some magnitude, and fairly rapid, the multicellular organisms do have a greater impact. This is not to say that the unicellular organisms are incapable of producing significant effects (historians will recall that the Black Death of medieval Europe and Asia was caused by unicellular microorganisms); we say only that the multicellular organisms produce more large effects and more rapidly. What this all adds up to is the conclusion that multicellular organisms are more successful than unicellular organisms, or that there are some definite advantages in being multicellular.

Why should it be advantageous to be multicellular? The answer seems to lie in the mathematical possibilities of combination. It is possible to have two or more cells get together and all do exactly the same things that they would do if they were separate. Many kinds of organisms, particularly among the bacteria, blue-green algae, and green algae, do just that. But the organism tends to be slightly more efficient if the cells achieve something analogous to cooperation. By cooperating the group of cells can either do more using the same amount of energy or do the same things using less energy. That is efficiency, and efficiency is highly advantageous.

Let's look at this from a rational, human viewpoint. (The cells, as far as we know, do not go through this rational review process to account for the results, but it can be helpful to us in understanding why things have happened as they have.) Every living entity is faced with a number of problems or activities that must be solved or performed in order to maintain that property we call life. All these activities require the use of energy and materials. Each and every living cell, existing as a separate organism, must perform all these activities and solve all these problems. Failure to do so results in death. Multiple-cell aggregations allow the possibility of dividing the required activities among different individual cells, which may then specialize in the performance of one activity, perhaps to the exclusion of any others. What results is an aggregation that collectively accomplishes all the features of life while its individual cells perform only some of the total number of necessities.

On the whole multicellularity tends to be more efficient than unicellularity in terms of materials and energy. The parallels with human social systems should be apparent.

The accomplishment of this division of labor is a bit more problematical. I think it is safe to assume that a cell surrounded by other cells is very likely to act differently from that same cell existing in isolation, because it will exist in a different environment. Irritability or sensitivity is one of the fundamental characteristics of life that was discussed in the first chapter. Living things respond or react to their environments. Different environments provide different kinds of stimulations, which, in turn, result in different kinds of reactions from the same kind of cell. A cell that is located in the center of a mass of other cells must exist in a different environment and be exposed to different stimuli from those of a cell located at the periphery of the same mass. Such cells should act in different ways, display different properties, and probably take on different appearances. Changes in the function and appearance of cells is called **differentiation.** Cellular differentiation modifies the environment of whatever cells are adjacent to those that have differentiated, which, in turn, promotes further differentiation. The net result can be an organism composed of many different kinds of cells all contributing to the functioning of the complete entity.

Furthermore, it should also be apparent that the individual cells of a multicellular organism are more insulated from the inhospitable aspects of the world than would be the case if they were living independently. A generalized, temporary environmental change affecting such factors as temperature, moisture, or acidity is apt to be more deleterious to individual, single cells than to the individual cells in a mass. The individual cells of a multicellular assemblage are each assisted by the existence of the other cells in the assemblage. Should some individual cells be lost or damaged, the chances are great that others will survive and continue to function, thus maintaining the integrity of the total organism.

The Hierarchy of Material Complexity

In examining the living and nonliving worlds, I have mentioned several things that can be considered as demonstrating different degrees of complexity. Organisms, whether composed of one cell or more than one cell, are quite complex.

SUBCELLULAR COMPLEXITIES

Even though cells are considered to be the fundamental unit of living organisms, we now know that cells are themselves composed of simpler constituents, such as membranes, organelles, and cytoplasm.

These cellular components consist of even more fundamental units of differing sizes and complexity called molecules. There are probably millions of different kinds of molecules. Molecules, in turn, consist of combinations of even smaller and simpler units called atoms, of which there are only 92 different kinds occurring naturally. Atoms are considered to be the fundamental units of **matter.** Depending upon how many and which kinds of atoms you combine, a nearly infinite number of molecules are conceivable. However, atomic physicists have learned that even atoms can be subdivided into subatomic particles, such as protons, neutrons, and electrons (plus about thirty other more esoteric entities). There appears to be a definite progression of increasing organizational complexity from the level of subatomic particles, through atoms and molecules, up to cells, which is paralleled by the subdivisions of the natural sciences. Subatomic particles and atoms are the levels of complexity that form the basis of physics. Chemistry deals only with those interactions at the levels of atoms and molecules. Biology is concerned with the higher levels of complexity that start with the cell.

ORGANISMAL COMPLEXITIES

Within the realm of biology the fundamental unit is usually considered to be the cell, although that does not preclude biologists from studying many of the subcellular levels of organization in an effort to understand the workings of cells. For all practical purposes, though, the basic unit of life is really the organism. For unicellular forms of life the two terms are equivalent; cells and organisms are the same thing (see Figure 2.20). But for multicellular forms of life there can be and are real differences between the organism and the cell. Multicellular assemblages can assume several different levels of organizational complexity depending upon degrees of differentiation (see Figure 2.21).

Colonies   It is possible, as alluded to earlier, for several independent, free-living cells to consistently gather together in a cohesive unit in which each of the constituent cells continues to function as if no other cells were present. This kind of arrangement is called a **colony** and is considered the simplest kind of multicellularity. It is frequently difficult to determine whether the colony or each individual cell is the organism (see Figure 2.22). If the colony is considered to be the organism, it is said to exhibit a cellular level of organization. Many forms of green algae, and possibly sponges, demonstrate this level of multicellular organization.

Tissues   True multicellular organisms exhibit a differentiation of cell types. The degree of differentiation shown can vary drastically. A con-

Figure 2.20

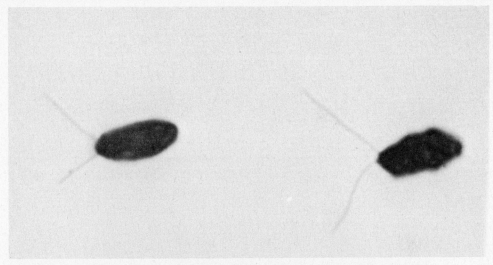

Two individual organisms called *Chlamydomonas*. As you can see, each individual consists of a single cell. Notice the pair of hairlike projections, called flagella. By rapidly beating these flagella in water, *Chlamydomonas* is moved from one location to another. [Photo courtesy of Carolina Biological Supply Company.]

tinuous mass of cells (a multicellular assemblage) in which all the cells appear similar and all perform the same functions is called a **tissue**. Some organisms, such as corals, sea anemones, and some brown algae, appear to have achieved the tissue level of organizational complexity. These organisms usually have several different kinds of tissues ar-

Figure 2.21

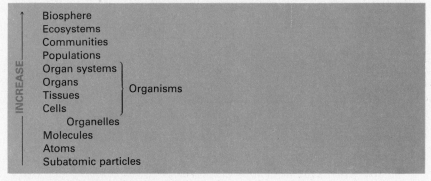

Levels of organizational complexity.

Figure 2.22

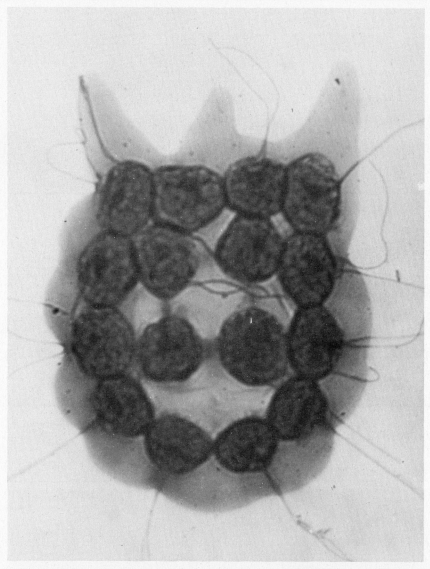

A simple colonial organism called *Platydorina.* Notice that the sixteen cells visible in this organism all look alike and that they all look a great deal like the individual cells of *Chlamydomonas* shown in Figure 2.20. At this level of organizational complexity decisions about the boundary between organism and cell may appear to be somewhat arbitrary. [Photo courtesy of Carolina Biological Supply Company.]

Figure 2.23

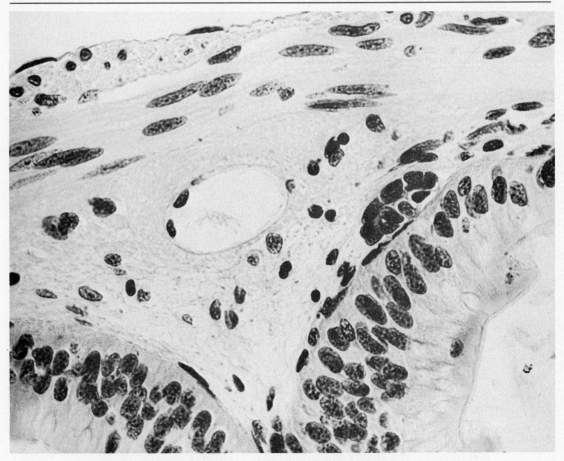

A cross section of a salamander intestine, magnified 620 times. Notice that there are four large groupings of cells, each grouping composed of cells that are highly similar. These groupings are called tissues. Notice further that the cell types found in one tissue are very different from those found in other tissues. [Photo courtesy of Dr. John B. Hess.]

ranged in some fashion that contributes to the integrated functioning of the total organism. They represent the simplest kind of division of labor. Examples of several kinds of tissues are found in Figure 2.23.

Organs   Different kinds of tissues may become massed together to form a multi-tissue assemblage that performs some one activity. Such an assemblage is called an **organ.** If you prefer, an organ can be thought

Figure 2.24

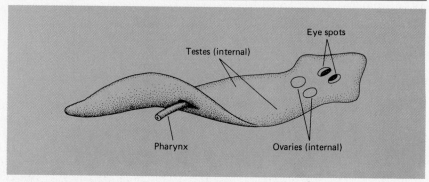

A freshwater flatworm, showing the organ level of complexity.

of as an assemblage of different tissues, all performing a specific function. There is a myriad of examples of organs with which you are all familiar: stomach, hand, lung, leaf, kidney, stem, and so on. Organisms that have differentiated to the degree of possessing organs are said to have achieved the organ level of organizational complexity. Flatworms and green plants are examples of these kinds of organisms (see Figure 2.24).

Organ Systems   Sometimes discrete organs may be united as a functional entity known as an **organ system.** An example of an organ system in humans is the digestive tract (or alimentary system), composed of the mouth, esophagus, stomach, small intestine, large intestine, anus, and associated glands such as the liver, pancreas, and so forth (see Figure 2.25). All these distinct organs function together in the process of taking food into the body of the organism. Organisms that have gained the organ-system level of complexity, as for example, the vertebrates and the arthropods, have several integrated organ systems.

The whole point of this discourse has been to show that in terms of organization some organisms are more complicated than others. There is a rough tendency (very rough) for the more complex organisms to become larger, but there are so many exceptions to this trend that it is an extremely risky generalization. For example, there are sponges organized at the cellular or tissue level that are physically larger than many worms organized at the organ level. As an even more extreme case, there are single-celled organisms such as *Acetabularia* that are larger than many kinds of multicellular insects that are organized at the organ-system level of complexity.

Figure 2.25

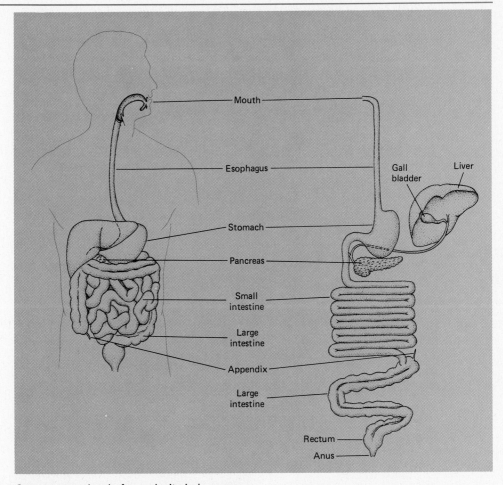

Organ system level of complexity in humans.

COMPLEXITIES ABOVE THE ORGANISM

Having learned that complicated organisms such as people are composed of organ systems, which are composed of organs, which are composed of tissues, which are composed of cells, which are composed of organelles, which are composed of molecules, which are composed of atoms, which are composed of subatomic particles, you probably feel like you are trying to solve the Chinese box puzzle. (You may also be wondering why you didn't study physics in the first place. At least there you only have to worry about two levels or organizational complexity.) But there are even larger systems of organization than these.

Populations   Organisms almost never exist as one-of-a-kind individuals. If you can name one kind of organism for which there is only one individual example, you will have named a kind of organism that is about to become extinct. Reproduction is a characteristic property of living organisms, which means that all organisms are derived from parents and will probably produce offspring. This is true for every kind of organism in existence and is a natural result of the cell theory. It implies that the various individuals within the kinds of organisms that I've been talking about are in some way related to one another ancestrally. Although it is not universally true, it is fairly common that two individual organisms of the same kind are required in order to accomplish reproduction. The complex of similar-appearing, and/or ancestrally related, individual organisms is called a **population** (see Figure 2.26). You can put whatever geographic limitations you want on a pop-

Figure 2.26

A rather simple example of a small population. These individual organisms are all members of the same species, are related ancestrally, and are capable of interbreeding to produce a new generation of rabbits. [Photo courtesy of Carolina Biological Supply Company.]

ulation (recognizing that they might be completely arbitrary), so long as you are only referring to one particular kind of organism. Thus it is possible to talk about the gypsy moths in your hometown, the Mockingbirds of Georgia, or the Gray Herons of Africa as perfectly good examples of populations. In the case of organisms that reproduce sexually, the population is a real entity in the sense that each individual member is potentially capable of interacting with any other member to produce new members of the population. Many biologists tend to use the terms *population* and **species** interchangeably. A species (note that the same spelling is used for both the singular and plural form) is a population of organisms that are capable, or potentially capable, of interbreeding.

Communities   Populations seldom, if ever, are found in isolation from other populations. Different species are almost always found in associations called **communities** (see Figure 2.27). These are not necessarily chance associations but are frequently functional interdependencies. Many species of organisms require the presence of other species in order to fulfill the requirements of life. These interspecific dependencies include the provision of foods, other materials, shelter, media for reproduction, maintenance of optimal population size, and so forth. The possibility of almost any species living without this vital interaction with other species is infinitely small.

Ecosystem   Communities must be very complex entities, based as they are on the interactions of their component populations. They are in actuality even more complex. Communities exist in the real, physical world and thus are composed not only of all the different living populations but also of all the nonliving materials and conditions that exist within some geographic context. The living components of the community must interact with one another and with the nonliving components of their surroundings. This system of interactions among living and nonliving materials of differing levels of organizational complexity is referred to as the **ecosystem.** It is worth emphasizing that both the living and the nonliving components of an ecosystem affect each other. Further elaboration of this idea will be found in subsequent chapters.

Biosphere   The ultimate extension of the ecosystem concept is the interactions of all local ecosystems on a global basis. All living things on the earth can be conceived of as making up one gigantic community that interacts with the nonliving components of the earth and is called the **biosphere.** The idea of some life form in the Amazon basin affecting living forms in the Siberian arctic may strike you as being highly im-

Figure 2.27

(a)

(b)

(c)   (d)

Different communities are commonly recognized by the most conspicuous plants found in them. Pictured here are some of the kinds of communities found in the world. (a) Grassland; (b) small pond; (c) deciduous forest; (d) tropical rain forest. [Photos (a) and (d) by the author. Photos (b) and (c) courtesy of John E. Wiley.]

Figure 2.28

This satellite photograph of the earth, taken from 22,300 miles (35,880.7 kilometers) above Brazil, illustrates the finite nature of our planet. Our normal perspective tends to encourage us to think of the planet as incredibly huge. All the life forms that we know anything about interrelate in a thin layer around the surface called the biosphere. [Photo courtesy of NASA.]

probable, until you take a look at some of the photographs that have been taken of our home planet from orbiting satellites or from the moon. From this new perspective (see Figure 2.28) one is struck by the isolation and finiteness of our planet. The only life forms that we know of are all found together on this one limited planet. They must interact globally to share the resources that occur nowhere else.

What I have tried to show in these last several pages is that we can perceive a continuum of organizational complexity within the material world (see Figure 2.21). As we move through this continuum, from atoms, through molecules and cells and species, to the biosphere, the number of units interacting with one another and the ways in which they can interact become extraordinarily large. Understanding the whole upper part of this continuum (from cell through biosphere) is the goal of biology. It is obviously an immense undertaking filled with complicated difficulties and has not resulted in many simple, universal generalizations. There are some generalizations, and the cell theory is one of them. Biology is a massive subject that is growing bigger every day.

Summary

1 / Many living forms exist that are too small to be seen directly by human beings. When the human visual sense is augmented with magnifying lenses, these living forms can be perceived.

2 / The use of microscopes has demonstrated that all living things have a cellular construction. The cell is the basic unit of structure and function of all living things.

3 / All living cells are derived from previously existing cells. Living cells cannot be produced through the interaction of noncellular or nonliving materials.

4 / A cell is composed of a watery fluid called cytoplasm surrounded by a semipermeable membrane.

5 / Eucaryotic cells have visible, membrane-bound subcompartments called organelles that specialize in the performance of various cellular activities. These compartments are not found in procaryotic cells.

6 / The major organelles found among eucaryotic cells are the nucleus, endoplasmic reticulum, mitochondrion, plastid, vacuole, and Golgi apparatus.

7 / While some living things occur in nature as a single cell, others are found as more or less integrated systems composed of many cells.

8 / When compared to single-celled organisms, multicellular organisms are much more diverse and appear to play a more prominent role in the world.

9 / Cellular differentiation allows for specialization of function and division of labor, processes that are generally more efficient in terms of materials and energy than the maintenance of uniform generalists.

10 / Organizational complexity increases from subatomic particles, through atoms, molecules, cells, tissues, organs, populations, communities, and ecosystems, to the biosphere.

11 / The material world, including living things, is organized with different degrees of complexity.

12 / Living systems are much more complex than nonliving systems.

Questions to Think About

1 / Of what importance to biology is the resolving power of the human eye? How does increasing the resolving power of the human eye through the use of a microscope add to our understanding of life?

2 / Why are all living things composed of one or more cells?

3 / The spoilage and/or decomposition of food materials like broth, meats, vegetables, and so forth have presented some very large problems to human beings for a very long time. How did the work of Pasteur help in solving these problems? How many different ways can you think of to preserve foodstuffs, and why do they work?

4 / Why do cells exhibit so many differences in shape and size?

5 / If the nucleus is responsible for controlling everything that a cell is and does, where does the nucleus get the necessary knowledge to perform?

6 / What does it mean for an organism to be successful?

7 / Why are there so many different kinds of organisms?

The Origin of Cellular Life

Introduction

As was discussed in Chapter 2, there is a well-established generalization in biology that all living organisms are composed of cells and that all living cells are derived from previously existing cells. The validity of this generalization has been demonstrated repeatedly, but it raises the question of ultimate causation. If living cells are all derived from previously existing cellular organisms, where did the previously existing cellular organisms come from? How far back in time can that generalization be used before various kinds of difficulties become apparent? If you adhere steadfastly to a literal interpretation of the cell theory, then eventually you are forced to conclude that cellular life as we know it has always existed. It is also necessary to assume that time is infinite. The concept of infinity, whether applied to numbers, time, space, or life, is one that frequently gives human beings difficulty. The idea that people exist along continua of time and space, which are so vast that they have neither beginnings nor ends, has a tendency to make many people feel uncomfortable.

This discomfort with infinity has been sidestepped in most cultures by devising narratives that are purported to explain both the origin of life on our planet and the origin of the planet itself. Figure 3.1 illustrates one of these culturally derived explanations. The important point about these stories, and there are hundreds of them, is that they all assume that life and the world are *not* infinite. The world had a definite starting point. Life also had a definite starting point, which occurred after the origin of the earth.

One of the beneficial aspects of the spontaneous generation hypothesis was that it did assume that life had an origin. Furthermore, it assumed that life could originate again and again and that it is still doing so. Repeated testing has demonstrated the fallacy of only this latter assumption; it offers no evidence on the validity of the original assumption. All that the tests indicate is that under the conditions specified it is not possible to originate living cells. What would happen to ideas about spontaneous generation if we assumed a set of conditions that are totally different from those that exist now? Let's assume, for example, that we are dealing with a brand-new planet on which life has never existed.

Conditions of Early Earth

Astronomers have concluded, from comparative studies of stars, that stars exist for finite periods of time having beginnings and ends. The time span involved in these stellar life cycles is in the tens of billions of

Figure 3.1

The ancient Egyptians believed that the world was formed when the Air-God Shu violently separated the Sky-Goddess Nut from her twin brother and husband, the Earth-God Geb. Geb's body convulsed into the earth's crust, while Nut's body arched over to form the starry sky.

years. Under some kinds of circumstances the materials of which the stars are made can also give rise to orbiting planets. There are, in fact, several hypotheses in existence to account for the origin of both stars and planets, the details of which need not concern us. What is noteworthy about all of these different hypotheses is their agreement that the planets would have had to be extremely hot at their inception (see Figure 3.2).

Geologists have been studying the structure of the earth for several hundred years now. Among the things that they have concluded about our planet is that its surface is largely composed of materials that have cooled from what must have previously been very high temperatures. An example of this observation is shown in Figure 3.3. These materials now form an outer crust surrounding a hot, semiliquid core that

Figure 3.2

Studies of huge galaxies of stars such as this one have led most astronomers to believe that planets revolving about individual stars must have been extremely hot at their inception. [Photo courtesy of Hale Observatories.]

still exists deep within the earth. The core material erupts to the surface occasionally in the form of volcanoes. The general idea is that very early in the earth's history the planet was a very hot, molten ball of material, very much like the material that now exists within the core.

Various means have been used to estimate the time at which the earth originated. The longer people have worked on these estimates, the older the earth becomes. Over the last two or three hundred years the estimates have increased by several orders of magnitude. At present the most widely accepted, reasonable estimate for the age of the planet earth is approximately $4\frac{1}{2}$ billion years (that is, 4,500,000,000 or 4.5×10^9 years). That is an extremely long time by human standards; for many of us 10 or 20 years ago seems like a long time. It would be extremely difficult to know much of anything about what happened several hundred years ago without recorded history. But if someone were to make a mammoth motion picture of the earth's total history,

Figure 3.3

Rock formations such as this one at Stone Mountain, North Carolina, reveal that much of the material of which the earth is composed was formerly heated to a molten condition. The central core of the earth is most probably still in this heated, molten state. [Photo courtesy of Matthew E. Stockard.]

and the movie ran for a total length of 24 hours, recorded human history would fill up only the last tenth of a second of the movie.

The early earth was certainly a lot different from the planet we occupy today. The surface temperature has been estimated to have been at least several thousand degrees Celsius. Most of this heat must have radiated into space. The gaseous atmosphere surrounding the earth would most probably have had a very different composition from the one present today. It probably consisted of substances that are chemically known as **reducing agents.** Simple examples of such substances that could have been ingredients in the primitive atmosphere include hydrogen (H_2), methane (CH_4), ammonia (NH_3), and water vapor (H_2O). The composition of gases emitted during the eruption of modern volcanoes confirms the idea that early earth conditions would have generated an atmosphere rich in simple reducing agents. The high surface temperature would almost certainly have generated strong convection currents in the atmosphere, resulting in violent storms. Hot gases are lighter than cold ones and consequently they rise, being replaced by cooler ones at higher altitudes. If water vapor was a part of the atmosphere, torrential rains would have accompanied the high winds, and spectacular displays of lightning would have occurred. It also seems probable that there would have been a strong influx of several kinds of radiation, including X rays, cosmic rays, and ultraviolet light.

I have presented here a reasonable hypothesis of what our earth was probably like at the beginning. It has received reasonable amounts of support from observations made in the present under presumably similar conditions. If you prefer, you can call this a complex supposition or an imaginary reconstruction.

What, if anything, does this description of the early earth have to do with life? You should have noticed that my description made no mention of living forms, and you can probably conclude from the physical circumstances that life (as we know it) could not have existed. Yet we know that now life does exist. These two points of reference define the boundaries within which our creativity may wander. How can we explain the existence of life? One possibility is that some living form, probably a **spore,** was transported here through space, found conditions suitable, germinated, and flourished. You can elaborate on this basic idea to the ultimate extreme and claim that all life on this planet was brought here from someplace else by some superrace of cosmic dwellers, but that really just avoids answering the question of how cellular life got started in the first place. Alternatively, we can hypothesize that the conditions described above can account for the spontaneous origin of living things.

Chemical Complexing

Living things are composed of unique chemical compounds. These compounds are found only in things that are, or have been, alive. Presumably if you are going to have living things, you have to have these organic compounds first, but that's another nasty paradox. You can't have organic compounds until you have living organisms to make them. Is it possible that the conditions described for the early earth could have led to the spontaneous production of chemical substances that today are unique to living things?

MILLER'S EXPERIMENTS

In the early 1950s an American graduate student named Stanley Miller sought an answer to just that question. Working in a chemistry laboratory at the University of Chicago, he built a functional model of the early earth based on the conditions outlined above. His model looked something like the contraption shown in Figure 3.4. The flame of the Bunsen burner represented the great heat input of the molten earth material. The atmosphere in the tubing consisted of methane, ammonia, hydrogen, and water vapor. The large chamber on the upper left had two electrodes, hooked up to a power source, that could generate sparks, simulating lightning. The condenser on the right duplicated the cooling effects of the upper atmosphere, which caused precipitation of liquids. The trap and large lower chamber represented reservoirs where liquids could temporarily accumulate as the surface slowly cooled.

Miller filled up his model world, turned it on, and let it run for a week. He then spent the next year or two trying to determine what was

Figure 3.4

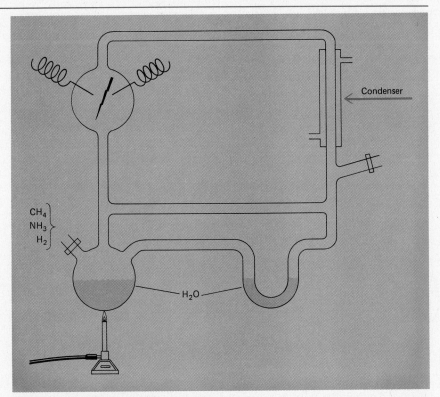

Miller's experimental apparatus.

now in his world. The results were startling. He found his world had produced 25 different chemical compounds that were thought to be associated only with living systems. These included twelve kinds of **amino acids** (fundamental units of proteins), urea (a common biological waste product), hydrogen cyanide, **acetic acid** (vinegar), and **lactic acid.** The reactions to Miller's findings were mixed. There were people who claimed his results were due to all kinds of artifacts, such as microbial contamination of the glassware, the water, or the gases. Maybe these chemicals were unintentionally introduced during the identification procedure.

The experiment was repeated many times by Miller and many others, taking all kinds of precautions against the possibility of accidentally introducing any of these chemicals into the experimental setup. In addition, several variations in the basic scheme were introduced, such as changing the composition of the reducing atmosphere, substituting various forms of radiation for lightning, and using different proportions of the atmospheric ingredients. The results were all essentially the same. All experimenters found that they could produce the kinds of molecules produced by living systems (organic compounds) if they set up the kinds of circumstances that were believed to be in existence during early earth history. The more they ran these model earths, the longer was the list of organic molecules they found, including **purines** and **pyrimidines** (basic constituents of nucleic acids), simple carbohydrates, and more kinds of amino acids. The amount of organic chemicals obtained was also impressive in that it accounted for 20 to 60 percent of the material harvested. The work of Miller and his followers demonstrated that our primitive world was very probably such that the production of a dilute soup of organic molecules could be expected to occur without the presence of living organisms. Or to put it another way, the **abiotic** (nonliving) production of organic chemicals was a highly probable event under the conditions that are presumed to have existed on the primordial earth.

POLYMERIZATION

The existence of an organic chemical soup does not qualify as the origin of life. Over the millions of years, as the earth's surface cooled and solidified, and as liquid water accumulated in surface depressions, organic molecules appeared in the waters. This same water, running over the surfaces of cooled rocks, dissolved some minerals and carried away some others in suspension, making an even more complex soup. Given enough time (which, as I have tried to point out, was abundantly available), it is possible for chance interactions to occur between individual organic molecules, which result in the production of larger molecules.

If there are localized areas with periodic flooding and evaporation, such as in present-day tidal pools, or local areas with a concentrated heat source, such as found around volcanic areas, then the probability of making large molecules increases. Very large molecules (called **macromolecules**) are distinctive features of living things.

The macromolecules of life are often of the type known as **polymers.** Polymers are gigantic molecules composed of repeating units of smaller molecules. For example, proteins are macromolecules composed of repeating units of amino acids; many carbohydrates are macromolecules composed of repeating units of simple **sugars;** nucleic acids are macromolecules composed of repeating units called **nucleotides,** which are in turn composed of a simple sugar, an inorganic **phosphate,** and either a purine or a pyrimidine. There are no living cells that lack polymers. It is possible to get simple organic compounds to abiotically polymerize into the kinds of complex organic compounds that we find associated with and characteristic of living forms today. It seems that natural physical conditions billions of years ago resulted in the production of many kinds of polymers without the involvement of living cells, although this does not normally happen today.

The existence of macromolecules still does not satisfy most people's requirements for life. While polymers and other macromolecules are invariably associated with life today, they do not constitute life all by themselves. What is necessary is to bring these macromolecules together as a functional unit that performs all the characteristic activities of life. In effect, it is necessary to construct a cell-like structure.

COACERVATION

It has been found that several kinds of molecules will form associations when placed together. These associations have been called **coacervates,** or coacervate mixtures, by the Russian chemist A. I. Oparin. The coacervates that have been examined are relatively simple, using water and other fairly large inorganic molecules. Coacervate mixtures start assuming features and characteristics that are very suggestive of different properties found in living cells. No one coacervate has exhibited all these features, but keep in mind that only simple coacervates have been examined. Coacervates might be an excellent model of what diverse mixtures of organic molecules can do under the proper conditions. They might even be a model of what happened in the primordial soup.

Coacervate mixtures usually form as isolated droplets or clusters. There may be many such clusters in a given amount of water. The cluster is often of colloidal dimensions so that it neither settles to the bottom of the water nor goes into true solution (see Figure 3.5). Such clus-

Figure 3.5

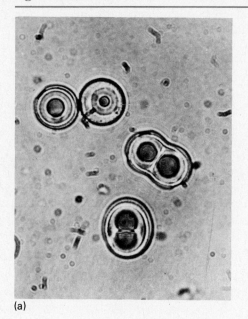

(a)

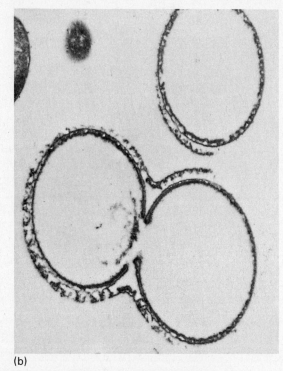

(b)

(a) Photomicrograph of protenoid microspheres. These little self-formed droplets are composed of mixtures of polymerized amino acids. They exhibit many of the properties described in the text for coacervates and hence are reasonable models for the origin of the first cells. These particular microspheres have been caught at different stages of division brought about by a slight increase in alkalinity. (b) Electron photomicrograph of a protenoid microsphere in the process of dividing. The major diameter of these droplets is about 3 micrometers. Notice in particular the complicated, multilayered structure of these droplets. [Photos courtesy of Dr. Sidney W. Fox.]

ters are frequently surrounded by a shell of water molecules, which are rigidly oriented relative to the cluster. This layer of water molecules forms a boundary that separates and defines the inside of the coacervate from its surroundings. The inside of the droplet is very different in composition and structural organization from the outside. Because the watery shell has a rigid orientation of its individual molecules (see Figure 3.6), it has a tendency to selectively attract and absorb some substances from the surrounding, randomly oriented watery medium, while ignoring or even repulsing others. These new molecules absorbed by the water shell may be incorporated into the structure of the coacervate droplet.

Figure 3.6

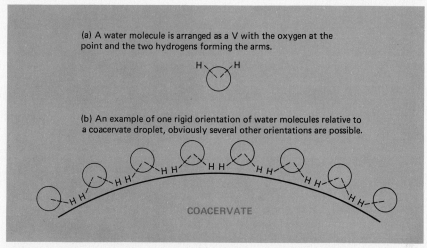

(a) A water molecule is arranged as a V with the oxygen at the point and the two hydrogens forming the arms.

(b) An example of one rigid orientation of water molecules relative to a coacervate droplet, obviously several other orientations are possible.

COACERVATE

Water molecules (H_2O) have a very special configuration of their constituent hydrogen and oxygen atoms.

Coacervate droplets tend to have a strong internal organization, which means that any new materials taken into the droplet must fit in without disrupting the basic structure. One of the ways this can be done is by arranging the new molecules immediately inside the water shell, making a more complex boundary structure. This arrangement provides even more selectivity of what passes in and out of the droplet. The boundary layer is becoming more membranelike. Alternatively, incoming material can be reorganized into additional interior material with the same strong internal configuration. Adding material to the droplet must result in a larger droplet. This increase in size by the droplet is directly comparable to the process of growth in living cells.

Because the coacervate droplet is both structurally organized and sharply separated from the external medium, the special conditions within the droplet will almost certainly exert a selective and regulative influence over the chemical reactions taking place within the droplet. In effect, the difference in conditions almost guarantees that what happens inside the droplet is different from what occurs outside the droplet. This situation gives a very crude description of metabolism.

Some coacervate droplets that are capable of increasing in size have been observed to become physically unstable. Once they reach some critical size, they have a strong tendency to fragment themselves. The result is several small coacervate droplets where originally there was only one large one. And here we have a crude model of reproduction or replication.

THE BOUNDARY?

It appears to be a safe assumption that physical conditions on the early earth would have promoted the production of a large number of organic molecules of varying complexity. It also seems highly probable that many of these organic chemicals could complex into systems very much like coacervate droplets. Because the number of different kinds of organic chemicals that could combine together to form coacervate mixtures is vast, it also seems safe to assume that a huge number of different kinds of coacervate droplets could have been formed. The particular kinds of molecules that were involved in the formation of the coacervate would have determined the properties and activities that were demonstrated by the coacervate. Among all the many possible coacervates, some must have been more dynamically stable than others. Such molecular complexes probably existed for a reasonably long period of time.

If by chance a droplet existed that had all the characteristics that I mentioned earlier—that is, a boundary layer that exhibited the crude properties of a membrane, a rudimentary form of metabolism, a primitive example of growth, and a simple form of reproduction—it would be tempting, and very convenient, to start using biological terminology (see Figure 3.7). Words such as *lineage, survival, parents, offspring,* and *natural selection* become almost unavoidable. Processes such as metabolism, growth, reproduction, and evolution are built into the very existence of this kind of coacervate droplet. If there exists a coacervate droplet that can grow and replicate, then there must also exist a **lineage** of coacervate droplets consisting of all those droplets that are derived from the ancestral droplet.

Those coacervate lineages that survived the longest must have been those that had the most harmoniously balanced structural and functional systems. Clearly this harmony had to exist not only in the metabolic and growth processes within the individual droplets but also in the replicative process that produced new generations of droplets. The only way a lineage could last for any number of generations would be if the offspring possessed all the traits and characteristics of their parent. Offspring that lacked one or more characteristics of the parent would be unable to perform all those functions that result in the attainment of the critical point where reproduction occurs. If reproduction does not occur, the lineage has come to an end. Clearly, then, any reproductive process that forms new droplets having the most faithful replication of the original droplet will maximize the survival of both

Figure 3.7

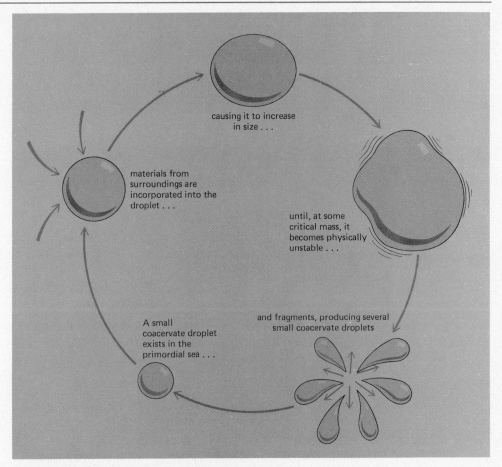

causing it to increase
in size . . .

materials from
surroundings are
incorporated into the
droplet . . .

until, at some
critical mass, it
becomes physically
unstable . . .

A small
coacervate droplet
exists in the
primordial sea . . .

and fragments, producing several
small coacervate droplets

The life cycle of a very complex, hypothetical coacervate.

the new droplets and the lineage. What I have described here are the very crude elements in a system of **natural selection** operating on a chemical level (see Figure 3.8).

There are very good reasons to think that a complex system of chemicals such as this probably existed a few billion years ago. The hard question to answer is whether or not such a system is living. The system appears to possess all the characteristics of living things that we recognize in contemporaneous forms, although many of the criteria are met in crude form. It would probably be unreasonable to expect the earliest forms of life to demonstrate all the refinements of the basic charac-

Figure 3.8

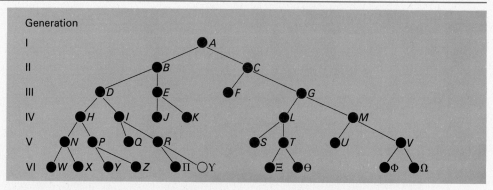

The concept of a lineage of protoorganisms, all tracing their ancestry back to *A*. Compare the success of the lineages starting at generation III. Because *F* leaves no offspring, its lineage terminates with itself. *E* leaves two offspring, which do not replicate, meaning its lineage terminates after one generation. If the success of a lineage is determined by the number of ultimately surviving offspring, which of the four individuals in generation III is the most successful?

teristics that today's life forms do. But the unwillingness of many people to accept these chemical blobs as living does serve to emphasize the extremely fuzzy separation between life and nonlife. The continuum of organizational complexity discussed in Chapter 2 reappears here as a continuum between nonliving chemical mixtures and unquestionably living cells.

If there is any connection by descent between these lifelike coacervate systems and living organisms of today, then it seems most probable that the complexed molecules involved in the composition of the coacervates must be the same large molecules that make up present-day cellular life, namely, nucleic acids, proteins, carbohydrates, and lipids. It would also make sense to assume that these substances functioned in a coacervate droplet many billion years ago in much the same way that they function in cells today. That may seem to be asking for too many results from the chance interaction of molecules, but when the vast spread of time is considered, it becomes, if not highly probable, at least less improbable.

Autotrophism

It has been implicit in the description above of the lifelike chemical systems that they were dependent upon the organic chemicals in the primordial seas for their nourishment. The warm soup, of which these

protoorganisms were a part, supplied all the materials needed to maintain their existence. In modern biological terminology such creatures are called **heterotrophic.** Heterotrophic organisms are those that require an external supply of organic chemicals as their source of both materials and usable energy. Human beings are examples of heterotrophic organisms, as are many kinds of bacteria. These protoorganisms must have been **anaerobic** also; that is, they obtained energy from the preformed organic compounds without using oxygen molecules. Such an assumption is required by the earlier description of the primordial atmosphere, which lacked molecular oxygen. Many kinds of contemporary bacteria are anaerobic, while others are **aerobic.** Humans, of course, are aerobic, requiring the presence of molecular oxygen.

An absolute dependence upon a finite supply of complicated organic compounds is a situation frought with hazards, for the simple reason that the supply of compounds acts as a limit on the protoorganisms. The continuation of the lineage requires the presence of those molecules that, to date, have only been produced by the chance interactions of physical factors. Presumably the more complex the molecule is, the more chance interactions are required to create it and, consequently, the more rare will be its occurrence. Sooner or later the protoorganisms' requirement for the compound will exceed the rate at which the abiotic world can produce that compound. At that point the protoorganism is on a dead-end street. To make matters worse, the more successful a lineage is at using a specific substance, the more quickly that substance will be removed from the surroundings, until the supply is virtually exhausted. Once the specific compound is gone, the continued existence of the protoorganism is jeopardized.

Let's assume, for the sake of an example, that there once was a protoorganism called a festue, which required a complicated organic compound represented by the symbols *ABCDEFGHIJ*. This complicated organic compound is composed of ten different, simpler organic compounds, each represented by a different letter, that have each reacted with one another in some fashion to form the larger molecule. The more festues there were, the more *ABCDEFGHIJ* was removed, making it more difficult to produce more festues. It also became more difficult to maintain those festues already in existence. The uptake and utilization of free-floating *ABCDEFGHIJ* by the festues can cause the elimination of the festues. Suppose that every once in a while the reproductive process of a festue ran a little bit awry, and instead of offspring that were perfect little replicates of the original festue, the offspring were slightly different in some way. It seems most probable that these imperfect copies would be defective and unable to function as a festue should. It also seems at least minimally possible that an oddball kind of offspring, which I'll call festue II, could have been produced that could take organic compounds *ABCDE* and *FGHIJ*, put them together forming *ABCDEFGHIJ*, and perform in every other way as though it were a standard festue. As the supply of naturally occurring *ABCDEFGHIJ* dwindled, the festue II could continue to function, which includes the production of festue II offspring, by using *ABCDE* and *FGHIJ*. The numbers of standard festues, because of the lack of *ABCDEFGHIJ*, must diminish (See Figure 3.9).

Using the same reasoning, we can expect the festue IIs to exhaust the supplies of *ABCDE* and *FGHIJ* and be largely replaced by festue IIIs, which synthesize *ABCDEFGHIJ* from still smaller constituents. Eventually, through the operation of the same processes, there could exist some festue XLVII that can assemble very simple and very abundant inorganic substances into the organic compounds that are then used to synthesize *ABCDEFGHIJ*. In the extreme such processes can produce a protoorganism capable of synthesizing all the organic compounds needed, using nothing but simple inorganic substances.

One point that may require clarification is the idea of reproducing oddball offspring. Offspring that possess new characteristics that can be passed on to the next generation are called **mutants,** and the process by which they arise is called **mutation.** This phenomenon has been studied extensively in modern organisms and almost invariably involves changes in the structure of nucleic acid molecules. It is an essentially accidental process, which occurs about once in every 100,000 cell replications. Such a phenomenon should not be unexpected in a lineage of protoorganisms.

A second point that requires elaboration is the idea of synthesizing large, complex organic molecules from simple, inorganic starting materials. This process is one that I've already shown the nonliving world did, but it required the input of a great deal of energy and a long duration of time. The time element is necessary because of the random nature of the interactions. The enforcement of some organization or direction on these interactions eliminates the necessity of the long time periods, but the energy requirement remains. Large, complex organic molecules are organized atoms, and the imposition of organization in a chaotic, random universe requires energy. Any protoorganism (or even any modern organism) that is going to organize atoms into complex molecules must make use of energy in the same way that Miller's model world made use of external energy sources.

In today's living organisms there are several sources of energy that are used. The most common source of energy is that released by breaking the chemical bonds in organic compounds. The most overwhelmingly common source of energy used to make the organic com-

Figure 3.9

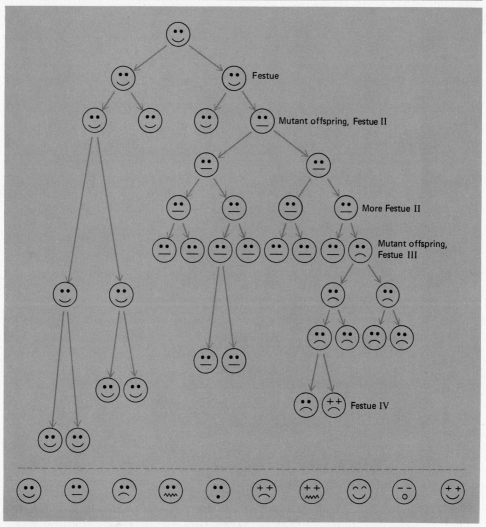

Origin of autotrophism. Festue requires *ABCDEFGHIJ* to grow and replicate. *ABC-DEFGHIJ* is a rare commodity, so festues grow and replicate slowly. Mutant offspring, festue II, can build *ABCDEFGHIJ* from more abundant and simpler *ABCDE* and *FGHIJ*. Festue II grows and replicates rapidly. More festue IIs deplete supplies of *ABCDE* and *FGHIJ*, slowing growth and reproduction. Mutant offspring, festue III, can build *ABC-DEFGHIJ* from *AB*, *CDE*, and *FGHIJ*, which are simpler and more abundant. Festue III becomes abundant. Festue IV is capable of making *ABCDEFGHIJ* from *AB*, *CDE*, *FG*, and *HIJ*.

Eventually there may be many kinds of festues, all having different nutritional requirements. One of these may be able to completely synthesize *ABCDEFGHIJ* from individual inorganic components such as *A, B, C, DE, F, G, HI,* and *J*.

Figure 3.10

A microfossil of an apparent procaryotic cell in the process of replication. It is one of many found in geological deposits in South Africa, which were laid down over 3 billion years ago. [From A. H. Knoll and E. S. Barghoorn, "Archean Microfossils Showing Cell Division from the Swaziland System of South Africa," *Science,* vol. 198, pp. 396–398, 28 October 1977. Copyright 1977 by the American Association for the Advancement of Science. Photo courtesy of Dr. E. S. Barghoorn.]

pounds from inorganic constituents in the first place is sunlight. The process by which this is accomplished is called **photosynthesis** and will be explained in more detail in Chapter 9. Organisms that are capable of synthesizing all their organic compounds from nothing but inorganic substances and an external energy source are called **autotrophic** organisms, or autotrophs. The existence of autotrophs guarantees a steady supply of the complex organic compounds needed by both the autotrophs and the heterotrophs.

Let me bring all this out of the realm of fantasy by giving a more straightforward description of the sort of creatures I have hypothesized into existence. These are small, single-celled, procaryotic organisms. Some of them are heterotrophic, others are autotrophic. Many of the autotrophs are photosynthetic. In fact, I am describing the group of organisms that I have previously called monerans. And the hypothesis begins to make even more sense when we realize that the oldest traces of living material found preserved in rock are of monerans (probably blue-green algae). These specimens are estimated to have been alive approximately 3 billion years ago (see Figure 3.10). Only by examining much younger rocks does one find evidence of more complex cells, and only in deposits of the last billion years (or less) does there appear evidence of multicellular organisms.

A Changed World

Thus it would appear that the Monera are of great importance and significance because they are the first living inhabitants of this planet. Their existence has had a major impact upon everything that has happened subsequently with respect to life. Their effects can probably be divided into three main areas: ancestry, heterotrophism, and photosynthesis.

ANCESTRY

If, indeed, procaryotic cells were the first living forms to inhabit our planet, then it seems most probable that all other forms of life have been derived from them. The modern monerans are living cells that have retained many of the structural characteristics of these ancestors, although they have undoubtedly acquired many new and different traits during the time of their existence. Eucaryotic organisms must be descendants that have developed new structural elements so that they appear to be totally different kinds of cells. However, this difference (or these differences) is more apparent than real, as has been mentioned in Chapter 2. I'll say more about the origin of eucaryotic cells later in this chapter.

HETEROTROPHISM

It seems most probable that these earliest life forms maintained their structural and functional integrity by consuming organic molecules from the surrounding medium. Organisms that do this today are called **decomposers** and are well represented among a group of monerans called bacteria. In a very real sense, decomposers function to eliminate any supply of organic molecules; they are a kind of cleanup squad. Originally the only organic molecules were those produced by the physical conditions of the nonliving world, but with the advent of primordial life forms, a new source of organic molecules was available. Impatient decomposers could consume and incorporate organic molecules from other organisms. This condition is called either **predation** (when the consumer is usually larger and rapidly devours the victim) or **parasitism** (when the consumer is usually smaller and slowly devours the victim). Examples of both of these life-styles can be found among the modern bacteria.

PHOTOSYNTHESIS

As mentioned earlier, the early life forms (at least some of them) must have acquired a means of trapping light energy from the sun in such a way that it could be used to synthesize complex organic molecules. Several forms of modern monerans, including the blue-green algae, perform this function today. The achievement of this skill establishes a kind of closed cycle, whereby the photosynthesizers produce organic molecules from inorganic materials and light energy, and the decomposers, predators, and parasites (as well as the producers themselves) consume these molecules, frequently releasing inorganic substances in the process. Photosynthesis also releases large quantities of molecular oxygen as a waste product.

"BRAVE NEW WORLD"

The world inhabited by these new life forms was decidedly different from what it had been. The most obvious difference was in surface temperature, but the living things themselves had caused a few changes. The oxygen released by photosynthesis had chemically altered the composition of the atmosphere. Oxygen is a reactive molecule, and large amounts of it would have converted the previous atmosphere into water, nitrogen gas (N_2), carbon dioxide (CO_2), and oxygen itself (O_2). Furthermore, at high altitudes oxygen is reorganized in small amounts to form ozone (O_3), which has the effect of shielding or blocking the passage of ultraviolet and cosmic radiation.

In brief, we find that the existence of living things, particularly photosynthetic organisms, has totally changed the physical conditions that led to the production of these living things. The circumstances that caused the origin of life have been eliminated by the living things. Duplication of the whole process is essentially impossible. The atmosphere is now composed of the wrong kinds of molecules, and almost all the original energy sources are greatly reduced. This situation makes abiotic (nonliving) production of organic chemicals highly improbable. But even if they could be produced, before they could be polymerized and complexed into a coacervate droplet, they would probably (almost certainly) either be destroyed by the oxygen in the atmosphere or consumed by some organism. It appears that living organisms have made spontaneous generation a dead issue.

Origin of Eucaryotes

While there are several obvious structural differences between procaryotic and eucaryotic cells, their chemical composition and the functions they perform are essentially the same. For this reason they are assumed to have a common ancestry. The chief difference between them is the presence of membrane-bound organelles in the eucaryotes. These organelles are really nothing more than subcellular compartments wherein related sets of chemical reactions occur. Organelles can be viewed as a means of subcellular division and specialization of labor. Thus it seems plausible that the eucaryotic condition is derived from the procaryotic condition.

ORIGIN OF THE NUCLEUS

Procaryotic cells do not have a nucleus, but they do have a supply of nucleic acid (DNA), which is the principal ingredient of the eucaryotic nucleus, and procaryotic DNA is often concentrated in a so-called nuclear area. In effect, these procaryotic cells do have a nucleus, but it

Figure 3.11

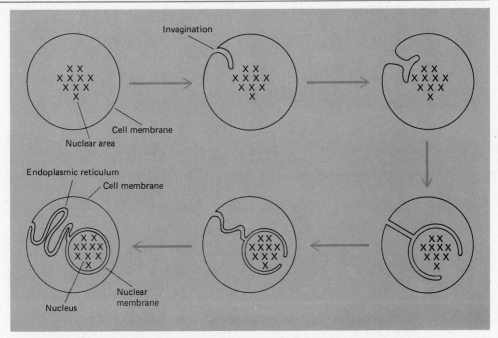

A proposed origin of the nucleus. It should be noted that no one has ever observed a cell membrane behave in this fashion.

lacks the surrounding double membrane. Because the chemical structures of cell membranes, nuclear membranes, and the endoplasmic reticulum (ER) are all the same, it is believed that the latter two might be derived from the former by some kind of **invagination,** as shown in Figure 3.11. It must be emphasized that this particular outline of events has never been observed in any living cell. While this explanation is not accepted by all biologists, it does satisfy the requirements of surrounding the nucleus with a double layer of membrane and connecting the outer nuclear membrane to the cell membrane by means of the ER. It is also probably more than coincidental that most of the functioning ribosomes are associated with the cell membrane in procaryotes and with the ER in eucaryotes. The invagination hypothesis accounts for this situation.

A simple invagination, such as illustrated in Figure 3.12, seems capable of explaining several kinds of organelles such as vacuoles. The formation of food vacuoles in protozoans has been observed to proceed in just this way.

Figure 3.12

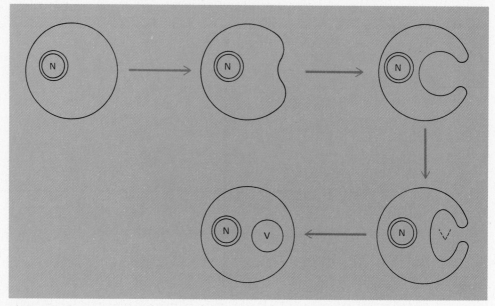

Origin of a vacuole.

PLASTIDS AND MITOCHONDRIA

Invagination can also be invoked to explain the origin of both mitochondria and plastids, but several peculiar features of each of these organelles argue in favor of a more complicated explanation. Both organelles can replicate themselves independently from the cell in which they are found. In addition, each carries its own supply of DNA, just as the nucleus carries the total cell's supply of DNA. It has been suggested that plastids and mitochondria might be derived from small procaryotic cells living inside larger cells (see Figure 3.13). This proposal has been called the *symbiotic origin* of the eucaryotes hypothesis because it is based upon the idea of two or more organisms living in association to the mutual benefit of each. Such a relationship has been called **symbiotic** or **mutualistic.**

In the case of plastids it is proposed that a small photosynthetic procaryote (perhaps like a blue-green alga) was engulfed by a larger heterotrophic cell, which was unable to digest the molecules of which the photosynthesizer was composed. Instead, the smaller cell continued to photosynthesize, producing an excess of organic molecules, which could be used as nourishment by the larger cell. The larger cell, having completely surrounded the autotroph, provided protection and mobil-

Figure 3.13

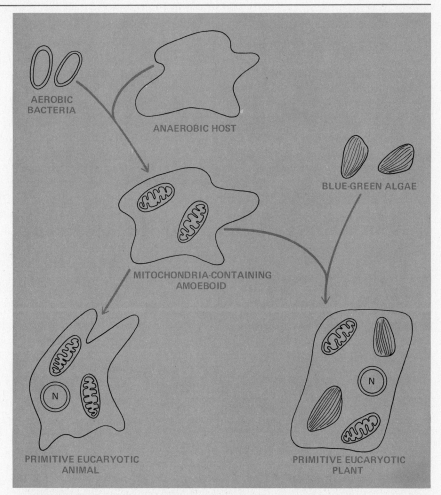

Some organelles are thought to have arisen through the development of an internal parasitic relationship between two kinds of procaryotic cells, a relationship that was mutually beneficial. [After Lynn Margulis, *Origin of Eukaryotic Cells,* New Haven, Yale University Press, 1970.]

ity. The longer this association lasted, the greater the mutual benefit and dependence became. Eventually, through the processes of mutation and selection described earlier, the smaller incorporated cell could carry out almost no functions other than photosynthesis, while the larger cell no longer hunted for food because the procaryote-plastid provided all the organic molecules needed.

The proposal for mitochondria works essentially the same way, except that the procaryotic ancestor of a mitochondrion is presumed to have been able to use molecular oxygen to extract energy from organic molecules. All organisms extract energy from organic molecules, but those that can use oxygen in doing so obtain a much higher energy yield and are called aerobic organisms. This aerobic energy extraction process, when it occurs in eucaryotic organisms, is only accomplished by the mitochondria. It is assumed that a procaryotic aerobe, engulfed by a larger cell but not destroyed, would degenerate in much the same way as the plastid-procaryote except for those activities concerned with aerobic energy release. This arrangement would be advantageous for both organisms involved in that the larger cell would benefit from the greater amount of energy released by the procaryote-mitochondrion, while the smaller cell would benefit from the larger cell procuring the organic molecules used in the energy-releasing reactions.

In this chapter I have tried to show that ordinary physical conditions can explain the existence of living cells on this planet. The occurrence of cellular life in recognizable forms is the natural result of the interaction of ordinary substances in processes that are quite understandable to modern people. This explanation may remove some of the aura of mystery surrounding the phenomenon of life; but, after all, that is just what science is supposed to do.

Summary

1 / The demonstrations that life cannot be originated under conditions that exist today imply that when life did originate, conditions differed from those of today.

2 / Observations made by astronomers and geologists suggest that early earth conditions were extremely hot and highly energized, with an atmosphere that was chemically reducing.

3 / Stanley Miller and subsequent workers have demonstrated that the proposed early earth conditions will naturally result in the production of simple organic compounds without the assistance of living material.

4 / Many chemists have described presumably natural conditions under which simple organic compounds may be polymerized abiotically.

5 / A. I. Oparin and subsequent workers have described a model system, called coacervates, in which complex, aggregated chemical mixtures display some of the characteristics of living systems.

6 / A hypothetical coacervate system composed of proteins, nucleic acids, and possibly carbohydrates and lipids functions in a way indistinguishable from that of the most primitive life forms.

7 / The earliest life forms were almost certainly heterotrophic, anaerobic, and procaryotic.

8 / Physical conditions and chemical laws would have favored the acquisition of autotrophic capabilities by some of the earliest life forms.

9 / *The presence of large numbers of photosynthetic autotrophs would most probably have released large amounts of molecular oxygen, thus drastically altering the composition of the atmosphere.*

10 / *The changed conditions of the world by the time both heterotrophic and photosynthetic organisms were present prevented the continued spontaneous origination of new life forms.*

11 / *Several compartmentalization schemes account for the existence of eucaryotic cells.*

Questions to Think About

1 / Suppose that at some future time people are able to engage in interstellar, or even intergalactic, travel. There is strong evidence that indicates that many stars have orbiting planets much as our sun does. Assuming that we could land on such planets, what kind of findings would convince us that these planets have their own forms of life?

2 / Why doesn't the dilute organic soup produced by Miller's experiment qualify as the origin of life?

3 / Why does the chemical composition of life have to be so complicated?

4 / Did the first living things on the earth have to be anaerobic heterotrophs? Why or why not?

5 / From what has been said in this chapter, autotrophs appear to have some distinct advantages over heterotrophs. Yet there are still (and it appears as though there have always been) a very large number of different kinds of heterotrophs. Why didn't all the heterotrophic forms get eliminated?

6 / What is the importance of the separate supply of DNA found in mitochondria and plastids?

Cellular Reproduction

Objectives

The student should be able to:

1 / *describe the process of binary fission.*
2 / *describe the process of mitosis.*
3 / *explain why the daughter cells from mitosis are genetically identical.*
4 / *explain why genetic identity may produce differences in appearance or behavior.*
5 / *construct a simplified sexual life cycle using the words* gamete, zygote, haploid, diploid, meiosis, *and* fertilization.
6 / *describe the process of meiosis.*
7 / *describe the similarities and differences between mitosis and meiosis.*

Introduction

I have already made several references to the subject of reproduction, and because it is an area of central concern to the science of biology, I feel it is necessary to give a more complete picture of this process. Reproduction is a subject of immense intrinsic interest accompanied by a remarkable volume of general ignorance. All living things reproduce themselves—so it is obviously a subject worth investigating in some depth as a means of understanding the basic characteristics of life.

BINARY FISSION

The form of reproduction that appears to be the simplest is that of **binary fission,** which is found in the monerans. In this process one bacterium simply constricts itself into two daughter cells, as shown in Figure 4.1. This process appears to be very much like the reproductive mechanism described earlier for the complicated coacervate systems involved in the origin of life. Those replicates (daughter cells) most closely resembling the constitution of the **antecedent** (parent cell) are most likely to survive and replicate themselves.

It follows, then, that binary fission must have some kind of organizing and controlling mechanism that guarantees (or at least maximizes the odds) that the resulting offspring will have the same constitution as the parent. Over the period of time since World War II a great deal of work has been done to clarify our understanding of the control mechanisms in moneran reproduction. It turns out that what appears on the surface as a very simple procedure is, in fact, extremely complicated. Rather than try to explain all these complexities, it is probably best to ignore the procaryotes and approach the more familiar, and apparently more complex, eucaryotic cells.

Figure 4.1

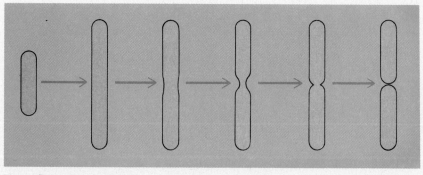

Binary fission in a bacterium.

Giving up the study of simpler organisms in order to study the same basic process in more complex organisms may seem paradoxical, but it's mostly a matter of convenience and technique. Procaryotic cells are very small, and they have no recognizable organelles within them to serve as landmarks. Consequently, it is very difficult to see what's going on inside a replicating procaryotic cell. In contrast eucaryotes are relatively large, and they do contain organelles, and it is possible to see all kinds of things happening inside the cell while it is replicating.

EUCARYOTIC REPLICATION

From earlier statements in Chapter 2 about the center for control of cellular function and replication, you would probably expect that the place to look for information about cellular reproduction is the nucleus. After biologists had looked at cells for a long time, they had built up a generalized scheme of what a eucaryotic cell was supposed to look like. One of the most critical parts of this scheme was a nucleus, bounded by a double membrane and containing chromatin.

This conceptual scheme got something of a rude shock in the mid-

Figure 4.2

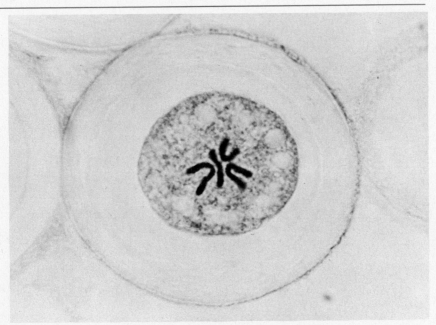

A eucaryotic cell that lacks a nucleus. Notice that it has, instead, four visible chromosomes. [Photo courtesy of Carolina Biological Supply Company.]

dle of the nineteenth century when some people found radically different-looking cells. Instead of finding a neatly organized nucleus in the
middle of a cell, they found a scattered bunch of little sausage-shaped
things that stained bright red in the same way as the nucleus (see Figure 4.2). These little bodies were called chromosomes (meaning "color-
ful bodies") and were searched for diligently. One of the first things
cellular microscopists noticed was that chromosomes couldn't always
be found, and when they were found they didn't always look the same.
Chromosomes are almost never found in some cells but are frequently
found in others. The most likely place to encounter chromosomes is in
cells where cellular replication is taking place. In multicellular organisms this means in areas of growth by cellular replication.

Studying cells with an ordinary light microscope necessitates
killing the cells and then staining them so that the insides become visible. By looking at huge numbers of these killed and stained cells in
which chromosomes could be found, it became apparent that a whole
sequence of events was occurring, which culminated in the production
of two new cells where there had been only one previously. This process is called **asexual** reproduction, because only one parental cell is
involved—there is no **sex** or exchange of **gametes** involved. The cellular mechanism by which asexual reproduction is accomplished is
called **mitosis.**

The process of mitosis consists of a sequence of events that until

Figure 4.3

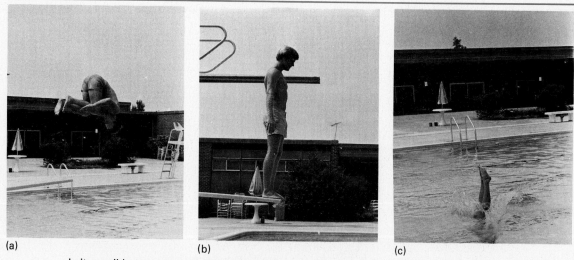

(a)   (b)   (c)

Is it possible to resequence these pictures in a way that is consistent with common
sense? [Photos courtesy of Matthew E. Stockard.]

relatively recently could only be perceived indirectly. The killing and staining of cells so that they can be seen through the light microscope automatically stops the process that is being studied. It is possible, however, to reconstruct the whole process if you have enough different instantaneous time samples of the process and if you know the beginning and the end of the process. The best analogy to this situation is one in which you have hundreds of individual moving picture frames of some relatively simple event, such as a person diving off a board into a swimming pool (see Figure 4.3). Knowing that the first frame must show the person at the back of the board and that the last frame will show toes protruding above the surface of the water, you can piece together all the frames in a coherent sequence that depicts the complete event. This indirect approach was the method first used to depict the process of mitosis. It is now possible, with the use of a technique called phase-contrast microscopy, to observe cell contents without killing and staining. This new technique has confirmed the validity of the description of mitosis formulated by the indirect approach.

Mitosis

PROCESS

It is not possible to overemphasize the fact that mitosis is a continuous process. For the sake of convenience and ease of communication, it has

(d)

(e)

been arbitrarily divided into several different stages. It should be clear that there will be nothing very solid about the boundaries between these stages, because the cells change very gradually from one stage to another. For example, for our analogy of the frames from a movie, it might be convenient to divide the frames into four stages: walking on the board, in the air with head up, in the air with head down, and in

Figure 4.4

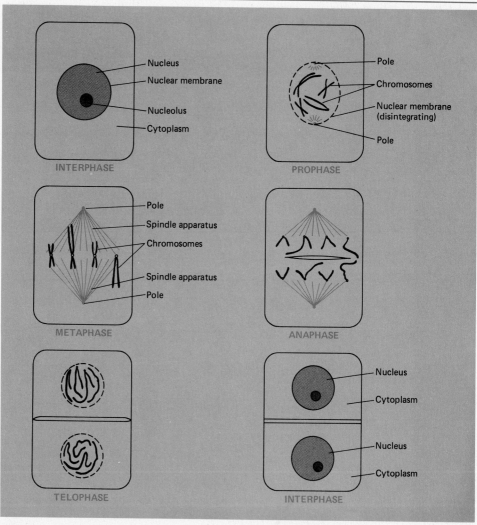

Mitosis.

contact with the water. These stages merge together sequentially, as do those of mitosis. With that precaution in mind, let me now outline the course of events that make up the process of mitosis (see Figure 4.4).

Interphase   A normal-looking eucaryotic cell that resembles the generalized description in Chapter 2 is a cell *not* undergoing mitosis. This is the appearance of cells before and after mitosis and is often called interphase of mitosis. The appearance of a cell in interphase might lead you to believe that it is reproductively resting, but this conclusion is not correct. In actuality, the interphase cell is unobtrusively doubling many of its cellular components in preparation for the more obvious events that occur during mitosis.

Prophase   The first stage of mitosis in which something appears to be happening is called prophase, and during this stage the cell's appearance is greatly altered. One of the first things to happen is that the chromatin, which is a diffuse fibrous material within the nucleus, starts to condense into discrete entities called chromosomes. Condensing in this case means that the long, thin nuclear material shortens and thickens greatly. Each chromosome can be seen to consist of two parallel strands of materials, called **chromatids,** connected together somewhere along their lengths by a structure called a **centromere** (see Figure 4.5). Because both chromatin and chromosomes react with Feulgen

Figure 4.5

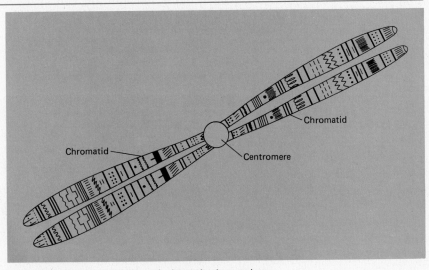

A chromosome as it appears during mitotic prophase.

stain to produce a characteristic red color, both are presumed to contain large amounts of the nucleic acid DNA. Every kind of organism tends to have a specific number of chromosomes that is characteristic for that organism. For example, human beings have 46 chromosomes; garden peas have 14.

While the chromosomes are shortening and thickening, the nucleus usually establishes **poles** at opposite ends. This action is accomplished in many different ways in different kinds of eucaryotic cells, but the net result is always that the nucleus has two opposite ends determined. At about the same time that the poles of the nucleus are established, the nuclear membrane appears to disintegrate. The double membrane surrounding the nucleus just disappears.

The chromosomes now go through a prolonged period of seemingly aimless drifting, during which a network of fine fibers appears extending from each nuclear pole to each centromere of a chromosome. Because this fiber network looks like two cones joined at their bases, it has been called a spindle apparatus or a **spindle.** With the establishment of a complete spindle, the chromosomes gradually drift toward a line midway between the two poles, the so-called **equator.**

Metaphase   The stage of mitosis in which all the chromosomes are momentarily lined up on the equator, with the spindle apparatus extending from each pole to each centromere is called metaphase. Metaphase is usually of extremely short duration.

Anaphase   As the mitotic process continues, the spindle fibers seem to shorten, while the centromeres duplicate themselves. This duplication of centromeres occurs in such a way that each chromatid of a chromosome has its own centromere, which is attached, by means of the spindle apparatus, to one of the two poles. As the spindle fibers shorten, each chromatid (now called a chromosome) migrates to one pole while its sister migrates to the opposite pole. This stage, called anaphase, guarantees that each pole receives one of each kind of chromosome. In the equatorial region, abandoned by the chromosomes, the cytoplasm appears to thin out in a number of different ways dependent on the kind of organism, a process called **cytokinesis.**

Telophase   The last active stage of mitosis, called telophase, is much like a reversal of prophase at each of the two poles with the addition of cytokinesis. Nuclear membranes re-form around each of the clusters of chromosomes, and the chromosomes themselves become greatly elongated and extremely thin, until they appear as normal diffuse chromatin. Cytokinesis continues with the formation of new cell membranes between the two new nuclei, resulting in two complete cells in interphase. The cytoplasm and its inclusions are often distributed about equally between the new daughter cells. As the chromosomes disappear from view in telophase, they are single-stranded; but when they reappear in the next prophase, they are once again double-stranded. Obviously some kind of duplication process is occurring during interphase, which makes it impossible to consider interphase as a resting stage.

IMPLICATIONS

Mitosis results in the production of two cells, each one of which has exactly the same complement of chromosomes as the original cells and thus is identical to its sister cell. Both new cells have the same supply of nuclear material or chromatin. Each cell has the same system of controls and regulations. Both cells have the same kinds and amounts of DNA. If we ignore the infrequent occasions of mutation, mitosis produces cells that have exactly duplicated nuclei. Both cells usually receive similar allotments of cytoplasmic material. In unicellular organisms repeated mitosis can result in the production of millions of individual one-celled organisms, all exactly alike. Alternatively, repeated mitosis can produce millions of identical cells in one mass, forming one large multicellular organism.

Except for a limited number of quite simple forms, multicellular organisms are not a mass of identical-appearing cells. They usually consist of many different cells, which neither look nor behave like each other. Different tissues and organs in the same organism are composed of very different kinds of cells. A rapid survey of your own bodies will reveal the presence of skin cells, hair cells, nerve cells, bone cells, muscle cells, and a multitude of others. Most of you are probably aware that multicellular organisms start out originally as a single cell. How can one cell produce all the different cells of a multicellular organism?

DEVELOPMENT

The changes that occur during the life of an organism are called **development.** The simplest example of development is probably aging. In multicellular organisms development includes the transition from a single cell to the production of many interdependent cells. The process of mitosis provides the necessary cells. A phenomenon known as differentiation enables the resulting cells to look and behave in different ways in spite of their all having the identical nuclear compositions that result from mitosis. This apparent contradiction can be explained by the interaction of at least three factors. One factor that can modify the effect of mitotic replication is an unequal division of the cytoplasm, which gives identical nuclei different systems with which to operate. A second factor known to affect the functioning of cells is the immediate environment in which they are found. A cell completely surrounded by other cells is in a very different kind of environment from a cell that has only one surface in contact with another cell. A third source of variability among cells produced by mitosis is that all the control and regulatory mechanisms within the nucleus do not have to function all the time. Cells with one set of controls in operation will look and/or behave differently from cells operating under a different set of controls.

It is assumed that despite the often radically different appearance and functioning of different cells within the body of a given organism, all the nuclei contain the same control and regulation information. This idea is given support by the observation that all the different kinds of cells within the same organism have the same number and kinds of chromosomes. Further support is derived from the observation that all cells in the organism contain the same amount and kind of DNA. In theory, then, each cell is potentially capable of replicating a complete new organism identical to the one from which the cell was removed.

Many kinds of organisms have a strong tendency to produce new organisms from parts of old organisms. Many kinds of plants, for example, are well known for their ability to generate replicates from cuttings. Starfish have been chopped into several pieces and each piece has regenerated a new starfish. There's a flatworm called a planarian that can be sliced in half, and each half will regenerate its opposite side (see Figure 4.6). To a lesser degree, salamanders can regenerate amputated legs and lizards can grow lost tails (see Figure 4.7). In human beings this ability is mostly restricted to the growing of new skin.

Individual, differentiated cells have been isolated from the roots of carrots and, when placed in the proper environmental conditions, have generated whole new carrot plants. Nuclei have been removed from the cells of swimming tadpoles and, when introduced into enucleated frog eggs, have directed the development of those eggs to mature frogs. Some people have proposed that this technique, called **cloning,** can be used to produce duplicate copies of individual people. Thus we could have an eternal supply of great people, such as J. S. Bach, Albert Einstein, Harriet Tubman, and Margaret Mead. However, reasonable doubts have been raised about the applicability of this technique on organisms as complex as human beings.

The conclusion that can be drawn from these observations and experiments is that mitosis is a very conservative process. Asexual reproduction, through mitosis, produces cells that all have the same potential as their immediate precursors. The information that passes from one generation of cells to the next (the so-called **genetic** or **hereditary**

Figure 4.6

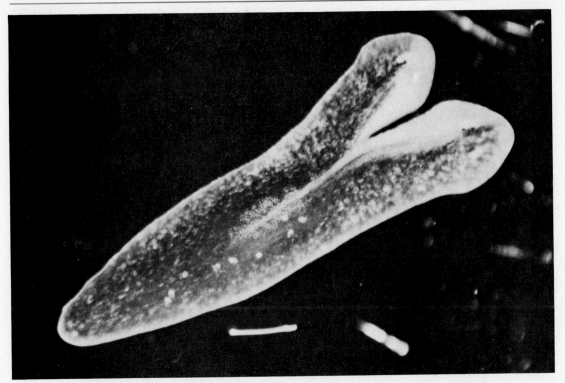

A planarian which has been partially split down its middle. Notice how each of the two sides of the animal is regenerating the missing half. The production the new cells needed to accomplish this feat is achieved by mitosis. [Photo courtesy of Carolina Biological Supply Company.]

information) is identical. Admittedly, the opportunity for mistakes in replicating the hereditary information (mutation) does exist, but mutation is of infrequent occurrence, and, on the whole, mitosis is quite dull.

The conservatism of mitosis is both its greatest asset and its biggest drawback. Living cells have found a scheme that faithfully reproduces new cells. These new cells can perform all the functions that were successful in their ancestors. Mutation allows the introduction of a little bit of variety, which can be retained if it works or rejected if it doesn't work. This asexual system of passing on genetic information works well. If the world had a very stable, uniform environment, asexual reproduction by mitosis might be completely satisfactory. But the world is not stable and not uniform. It is changing constantly. Biologi-

Figure 4.7

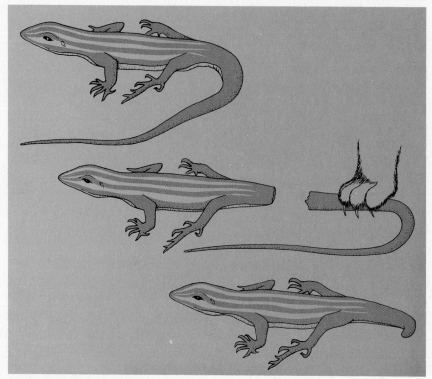

Regeneration of a lost tail.

cal solutions that work successfully under one set of circumstances may not succeed if those conditions are altered. Under such fluctuating conditions as exist in the real world, it is more advantageous to have a great deal of variety and diversity among offspring. Differing offspring maximize the chances that some will survive in the new regime. The question now becomes one of how to maximize the amount of variation within the offspring.

Sexual Reproduction: Meiosis

The process whereby many living organisms ensure diversity of genetic information among their offspring is called sexual reproduction. **Sexual reproduction** involves the fusion of two specialized cells, called **gametes,** to produce a new individual cell, called a **zygote.** The zygote contains an equal representation of hereditary information from each of its two parents; thus it is identical to neither.

Because the zygote contains the genetic information from both gametes, it has twice as much hereditary material as either of the two gametes involved in its formation. Conversely, each gamete contains half the amount of hereditary information found in a zygote. In eucaryotic organisms it is known that this hereditary information is packaged in chromosomes. We can conclude, therefore, that a zygote containing eight chromosomes was formed by the fusion of two gametes, each bearing four chromosomes. This difference in chromosome number (or genetic information) between gametes and zygotes is indicated by the use of some new words. The double number of chromosomes as found in zygotes is called the **diploid** condition, and hence zygotes are diploid cells. Gametes, containing half the number of chromosomes found in zygotes, are called **haploid** cells, or exhibit the haploid condition. The fusion of two haploid gametes will produce one diploid zygote. The haploid configuration is often abbreviated to the letter N, the value of N being equal to the number of chromosomes. Diploid cells then have a value of $2N$, because they are derived from a cell produced by the fusion of two cells, each with a value of N.

It's easy to see how cell fusion, called **fertilization,** produces one diploid cell from two haploid cells, but where did the two haploid cells come from? That's the other half of sexual reproduction. There is a special variation of mitosis, called **meiosis,** that occurs only in certain cells. In multicellular organisms these gamete-producing cells are usually clustered together in organs called **gonads.** Diploid cells within the gonads undergo meiosis to produce haploid gametes. Frequently there are two different kinds of gametes produced: male gametes and female gametes. Sometimes the process of meiosis is called **gametogenesis.** Sexual reproduction requires both meiosis and fertilization.

PROCESS

Like mitosis, meiosis is a sequential process that has been arbitrarily divided into a number of stages. Keep in mind these stages are purely for convenience and ease of communication in relating all the different phenomena that occur during the full process. Many people are inclined to think of meiosis as two processes; they are wrong. The process of meiosis is illustrated in Figure 4.8.

Interphase   Meiosis starts with a cell that looks just like any ordinary eucaryotic cell, so it is said to be in the interphase. This terminology is the same as that used in mitosis, which could be confusing. But since

Figure 4.8

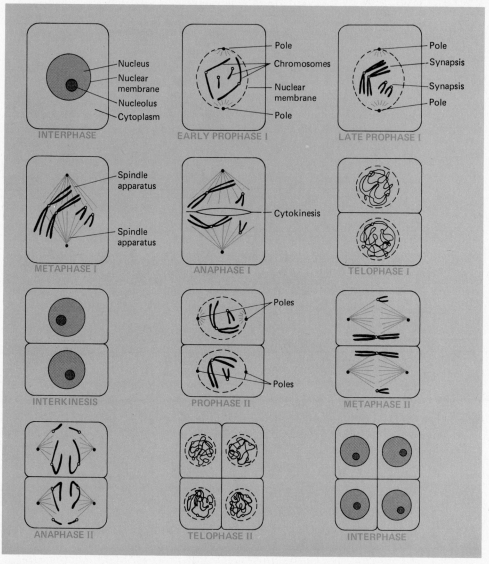

Meiosis.

there is no way of telling the two states apart, it seems useless to invent a new term.

Prophase I   The active, visible part of meiosis starts out in much the same way as does mitosis. The chromatin starts to condense into chromosomes, but, unlike the chromosomes in mitosis, these chromosomes initially appear to be single-stranded. Each chromosome consists of a single chromatid with a centromere somewhere along its length. Nuclear polarity is established, and the nuclear membrane disintegrates. The fully condensed chromosomes (still single-stranded) now do something very unusual. Each chromosome pairs up with another chromosome that looks just like it.

These two chromosomes are the same length, have their centromeres in the same relative position, and have the same elaborate pattern of banding. Such highly similar chromosomes are said to be **homologous.** One of the homologous chromosomes was originally contributed to this diploid cell by the paternal gamete, while the other was contributed by the maternal gamete. Diploid cells not only have twice as many chromosomes as do haploid cells, but they have two of each kind of chromosome possessed by a haploid cell. The easiest way to define a diploid cell is to say that it is a cell that has two of each kind of chromosome (see Figure 4.9). Haploid cells only have one of each kind of chromosome.

Homologous chromosomes come together to form synaptic pairs, or a **synapsis,** and almost immediately each chromosome assumes the double-stranded condition. Each chromosome now consists of two parallel chromatids connected by a centromere. Thus the synaptic pairs appear as four parallel strands of chromatids, sometimes called **tetrads.** Each synaptic pair or tetrad goes through a period of drifting while the spindle apparatus is set up, eventually moving toward the equator.

Metaphase I   During metaphase I each synaptic pair of homologous chromosomes lines up on the equator. The spindle fibers extend from each pole to each synaptic pair, but it would appear that only one spindle fiber attaches to each of the two centromeres. As in mitosis, this stage is of very short duration.

Anaphase I   In this stage of meiosis the spindle fibers shorten and the homologous chromosomes are separated from their synapses. One member of each homologous pair then migrates toward each pole. In contrast to the situation in anaphase of mitosis, the centromeres do not replicate in anaphase I of meiosis. Each chromosome at each pole is still in the double-stranded condition. But as in anaphase of mitosis, cytokinesis is initiated at the equator.

Telophase I   This stage of meiosis initiates the establishment of a nucleus at each pole. The chromosomes might be said to uncondense (each chromosome thins and elongates into chromatin), the spindle ap-

Figure 4.9

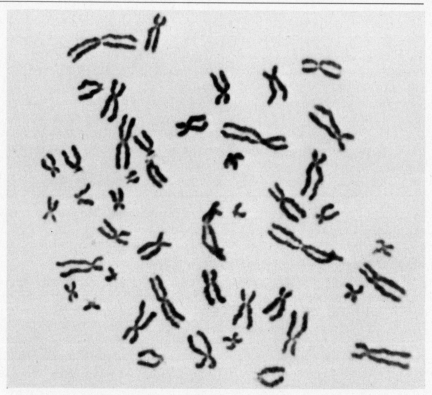

Chromosomes prepared from a normal human female. Notice that there are 46 chromosomes. If you examine closely, you will discover that there are only 23 kinds of chromosomes with 2 copies of each kind. [Photo courtesy of Carolina Biological Supply Company.]

paratus breaks down, nuclear membranes re-form, and cytokinesis continues to completion.

Interkinesis   Once telophase I is completed, both cells go into a short resting stage called interkinesis. The two cells resemble ordinary cells in interphase but are still in the process of meiosis. Interkinesis is a convenient marker for separating meiosis into two halves. During interkinesis both nuclei have one of each kind of chromosome in the double-stranded condition, but these are not visible.

Prophase II   In this stage activity resumes as the chromosomes of each nucleus once again condense. Each chromosome consists of two chromatids arranged in parallel and connected by a centromere. There is one of each kind of chromosome in each of the nuclei. Polarity is established in each nucleus, and the nuclear membranes break down. The chromosomes drift around as the spindle apparatus forms. Spindle fibers from each pole attach to each centromere, and both sets of chromosomes move toward their respective equators. Prophase II is thus exactly like a mitotic prophase except that it is happening in two adjacent cells simultaneously.

Metaphase II   In this phase each set of double-stranded chromosomes lines up at the equator of each of the two cells. The spindle apparatus in each cell extends from both poles to the centromere of each chromosome.

Anaphase II   Each centromere of each chromosome replicates itself in this stage so that each chromatid now has its own centromere and is thus called a chromosome. The spindle fibers shorten, and one of each kind of chromosome moves toward each pole. Two sets of cytokinesis are initiated, one at each equator.

Telophase II   Finally, at each of the four poles the chromosomes start to uncondense, the spindle apparatus breaks down, nuclear membranes are re-formed, and cytokinesis is completed. Again, this phase is just like mitotic telophase, except that twice as many cells are involved.

Thus meiosis starts with one diploid cell and ends with four haploid cells. These haploid cells may or may not function immediately as gametes.

Mitosis and Meiosis in Perspective

Descriptions of the mechanical operations of mitosis and meiosis are very likely to engender panic among students. Admittedly there are many new words that have been used to describe some very precise phenomena, and some of these will have to be memorized. However, it is much more important to achieve an understanding of what is occurring in each of these processes, how they are similar, and how they differ.

Mitosis is really just a means of parceling out equal amounts of nuclear material to the daughter cells. The process ensures that each resulting cell has the same number and kind of chromosomes as the mother cell. It stands to reason that both haploid and diploid cells can undergo mitosis, diploid cells producing more diploid cells and haploid cells producing more haploid cells.

Meiosis is a means of reducing the number of chromosomes in the cell by half, and thus it is only possible in cells that are initially diploid. Meiosis starts with one diploid cell and ends with the production of four haploid cells. The only normal way known to reverse this reduction of chromosome numbers—that is, to go from the haploid condition to the diploid condition—is to fuse two haploid cells in the process of fertilization.

Probably the easiest way to see how these processes fit together is to examine rough outlines of the life cycles of some organisms. Life cycles demonstrate the different ways these processes are used. The particular organisms chosen are merely illustrative and have no particular significance to the processes of mitosis or meiosis.

SPIROGYRA

Spirogyra is a filamentous green alga, commonly found in fresh waters around the United States. It is a photosynthetic eucaryote, composed of essentially single cells arranged in a line. Each cell in the filament is considered to be haploid. Each cell in the filament is able to undergo mitosis, thus making the filament longer. Sexual reproduction usually occurs in the autumn, before the start of truly adverse weather. Cells from two different filaments, lying side by side, start to bulge toward each other. This bulging continues until they meet and fuse, forming a tube between the two cells. The contents of one cell then travel through the tube to the second cell, where the nuclei fuse. This process produces a diploid zygote, which synthesizes a very thick wall around the cell. This thickened cell wall apparently helps the zygote to withstand the rigors of winter when the haploid cells die. In the spring the zygote undergoes meiosis, producing four haploid cells. Three of these die, and the one remaining grows through the cell wall, undergoes repeated mitotic replications, and generates a new filament of *Spirogyra* (see Figure 4.10).

MOSS

Mosses are small green plants that occur commonly in moist areas. They are multicellular, photosynthetic eucaryotes. Moss plants originate as a single haploid cell, which replicates by mitosis to produce a horizontal filamentous structure called a **protonema** (see Figure 4.11). Further mitosis produces downward-directed filaments, called rhizoids, which enter the soil, and upward-directed, aerial structures, which we recognize as the moss plant. All these cells are haploid because they are the result of mitosis. At the top of the moss plant certain groups of cells differentiate as gonadal structures of two different kinds, **antheridia** and **archegonia**. Antheridia and archegonia may de-

Figure 4.10

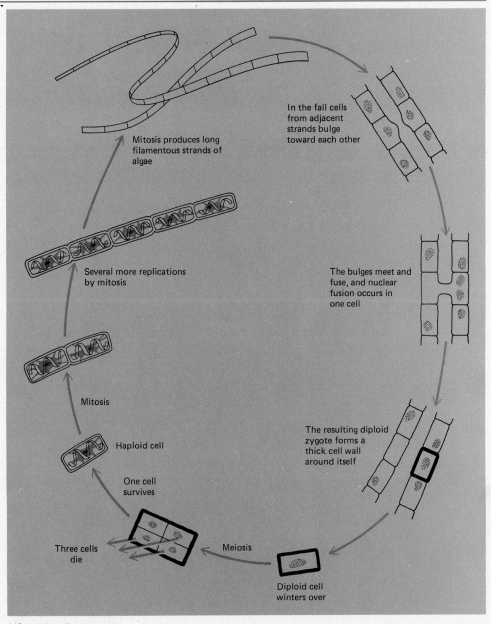

In the fall cells from adjacent strands bulge toward each other

Mitosis produces long filamentous strands of algae

The bulges meet and fuse, and nuclear fusion occurs in one cell

Several more replications by mitosis

The resulting diploid zygote forms a thick cell wall around itself

Mitosis

Haploid cell

One cell survives

Three cells die

Meiosis

Diploid cell winters over

Life cycle of *spirogyra.*

Figure 4.11

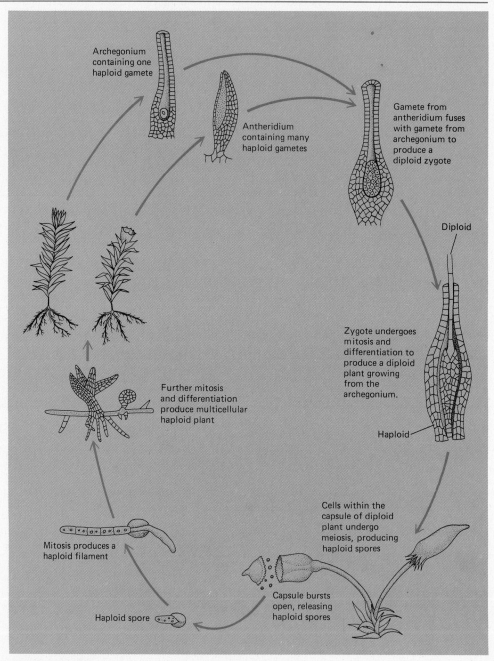

Life cycle of moss.

velop in different plants or the same plant, depending upon the species of moss.

The archegonium is a multicellular, flask-shaped structure containing one large haploid gamete in its core. The antheridium is a tubular sack containing a large number of small **flagellated** haploid gametes. During rainstorms, or even during a heavy dew, the antheridia open up, releasing the enclosed cells, which then swim through the surface layer of water to the archegonia. Locomotion of these gametes is accomplished by beating their whiplike flagella through the water, and the archegonia are apparently located by following a concentration of a chemical substance emitted by the archegonium. Fertilization occurs within the archegonium when the nucleus of the swimming gamete fuses with that of the large gamete lying at the core. Thus a zygote is produced.

The zygote then undergoes repeated mitotic replications, producing a multicellular stalklike structure that grows out of the archegonium. The top of this stalk expands to form a multicellular capsule. All the cells derived from the zygote are diploid, having two of each kind of chromosome, one from each of the two fertilizing gametes. Within the capsule there are many special cells that undergo meiosis, each producing four haploid cells called spores. Changes in humidity cause the capsule to open, releasing these spores to the wind, which carries them away. Those haploid spores that land in a suitable place undergo mitosis and germinate, forming a new moss protonema.

PINE

Pines are large trees that occur commonly throughout the Northern Hemisphere. Like mosses, they are multicellular, photosynthetic eucaryotes; but, in contrast to the situation for mosses, the familiar, conspicuous part of the plant is diploid. Pine trees grow by mitosis from tiny diploid **embryos** found in pine seeds. It may require as many as thirty years or more for pines to reach maturity and to be capable of sexual reproduction. Until this time all production of new cells is the result of mitosis. At maturity special organs differentiate, which, again, are of two kinds: **staminate** cones and **ovulate** cones. The familiar, woody pinecones are mature ovulate cones that are a year or more old (see Figure 4.12).

Within the ovulate cone are diploid cells that undergo meiosis to produce four haploid cells. One of these four haploid cells then undergoes mitosis, producing a multicellular, haploid structure within which are several archegonia. Each archegonium contains one large gamete called an **egg.** Keep in mind that this whole haploid structure is almost completely enclosed in diploid cells of the ovulate cone, the

Figure 4.12

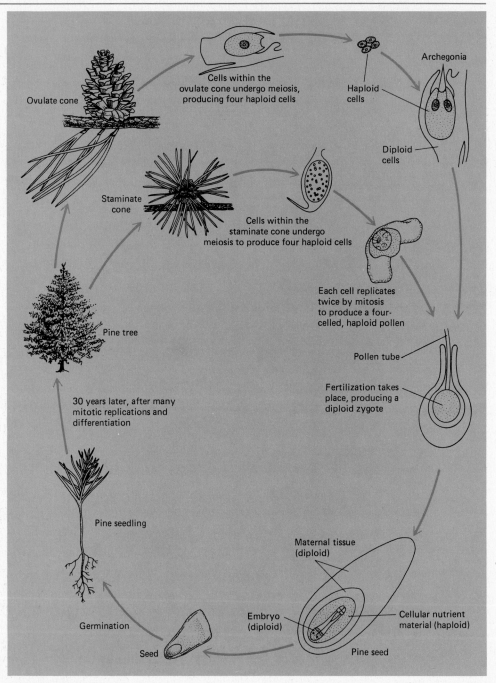

Cells within the ovulate cone undergo meiosis, producing four haploid cells

Archegonia

Haploid cells

Diploid cells

Ovulate cone

Staminate cone

Cells within the staminate cone undergo meiosis to produce four haploid cells

Each cell replicates twice by mitosis to produce a four-celled, haploid pollen

Pine tree

Pollen tube

Fertilization takes place, producing a diploid zygote

30 years later, after many mitotic replications and differentiation

Pine seedling

Maternal tissue (diploid)

Embryo (diploid)

Cellular nutrient material (haploid)

Germination

Seed

Pine seed

Life cycle of pine.

whole structure being called an **ovule.** There are usually many individual ovules within each ovulate cone.

Within the staminate cone there are also diploid cells that undergo meiosis to produce four haploid cells. Each of these haploid cells then undergoes two mitotic events, producing a four-celled, haploid **pollen** grain. The pollen grains are released from the staminate cones and carried by the wind. If and when a pollen grain lands on an ovulate cone, it grows a long extension (called a pollen tube) down into an archegonium. The pollen tube allows passage of a haploid pollen nucleus to the egg, where fertilization takes place.

The resulting diploid zygote undergoes several mitotic replications, forming a diploid embryo. The remaining haploid cells within the ovule replicate and differentiate into material that will be used to nourish the developing embryo. The diploid cells surrounding the ovule also replicate and differentiate to form a skinlike covering, which is extended at one end to form a wing. This whole structure of diploid embryo, haploid food material, and diploid skin with wing requires approximately a year of development within the ovulate cone before it is released as a fully formed **seed.** The seeds are then dispersed, and those that settle in a suitable location undergo mitosis within the embryo, resulting in a new pine seedling and a new diploid generation.

HUMAN BEINGS

Human beings are large, heterotrophic, eucaryotic animals that occur all over the world. They are diploid and occur in two forms, male and female. Within the males are paired organs called **testes** in which specialized cells undergo meiosis, producing four haploid cells. Each of these haploid cells then undergoes several mitotic replications and differentiation, resulting in a large number of haploid, flagellated cells called **sperm.** Within the females there are paired organs called **ovaries** in which specialized cells undergo meiosis to produce four haploid cells. Three of these cells, which, because of unequal cytokinesis, receive very little cytoplasm, are called **polar bodies** and quickly degenerate. The fourth haploid cell has essentially all the cytoplasm and is called an egg or **ovum.**

An egg is released from an ovary periodically and allowed to leave the body through a series of tubes known as the female reproductive tract. If sperm, transferred from a male to a female's reproductive tract, should encounter one of these eggs, fertilization results and a diploid zygote is formed (see Figure 4.13). The zygote then implants itself on the wall of part of the female reproductive tract and undergoes a long period of mitotic replication and differentiation. At the end of this period a new multicellular, diploid person is released from the female's

Figure 4.13

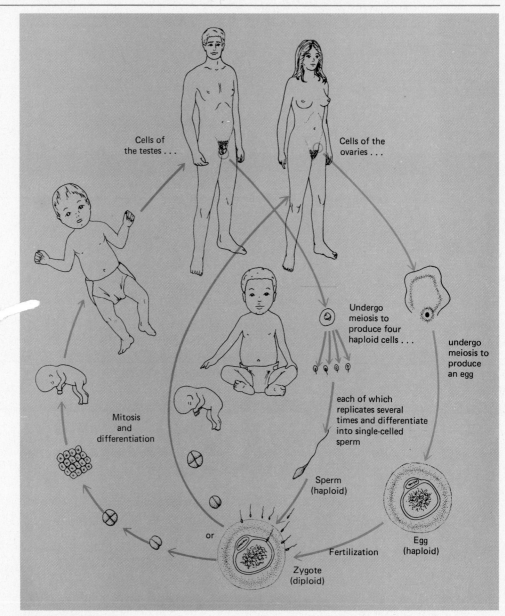

Life cycle of humans.

reproductive tract, and the male responsible for fertilization is supposed to give all his friends and acquaintances a cigar.

Summary

1 / Binary fission is a simple means of asexual reproduction found in procaryotic organisms.

2 / Asexual reproduction in eucaryotic cells is called mitosis.

3 / Mitosis involves the equal replication and partitioning of chromosomes to two daughter cells. Thus the two cells resulting from mitosis contain the same genetic material, the same system of controls and regulations, and the same kinds and amounts of DNA.

4 / In multicellular organisms cellular differentiation may produce cells that appear and function differently, although all contain the same genetic information.

5 / Fertilization is the fusion of two haploid gametes to produce a diploid zygote.

6 / Haploid cells contain one of each kind of chromosome.

7 / Diploid cells contain two of each kind of chromosome.

8 / Meiosis is a process of cellular replication in which one diploid cell produces four haploid cells.

9 / Sexual reproduction involves the alternation of meiosis and fertilization, meiosis and fertilization, and so on.

Questions to Think About

1 / Why should all the cells of a human body (except the gametes) contain the same number of chromosomes?

2 / Is there any particular reason why the diploid chromosome number for people must be 46?

3 / If there were some other kind of animal that also had a diploid chromosome number of 46, would that automatically make it possible for such an animal to sexually reproduce with a person?

4 / What is a chromosome?

5 / In what parts of the human body would you expect to find the greatest amounts of mitosis?

6 / In what parts of the human body would you expect to find the greatest amounts of meiosis?

7 / How do mitosis and meiosis differ from each other?

8 / In what ways are meiosis and mitosis the same?

9 / Why are there so many similarities between meiosis and mitosis?

5

Heredity

Introduction

Chapter 4 introduced you to the mechanics of reproduction at the cellular level. You have already seen in Chapter 2 that what the cell is and does is determined by its nucleus. The combined characteristics of cells determine the properties that are exhibited by multicellular organisms. How the regulatory information of the nucleus is transmitted from one generation to another is thus of fundamental importance in understanding reproduction.

For most people the idea that the nucleus or the chromosomes are the repository of cellular information is a vague abstraction with no obvious connection to normal reproductive experience. It can be demonstrated that the idea is probably correct, but it has no relevance. What is needed at this point is a way of relating the notion of nuclear control material to the sorts of things with which we are familiar.

LIMITATIONS OF HEREDITY

One of the most basic and obvious aspects of reproduction, whether asexual or sexual reproduction, is that the offspring that result are the same species as the parent(s). Chickens, for example, don't give birth to tigers or eagles or even turkeys. When chickens reproduce, they make more chickens. This very simple idea implies that there is something about the biological makeup of a chicken that enables it to pass these characteristics of chickenhood onto its offspring. The part of biology that deals with the transmission of characteristic properties from one generation to another is called **genetics** or **heredity.** The basic facts leading to an understanding of heredity must be derived from observing the results of reproduction within a given species or, by comparison, within several species. Although there have been many uproariously funny jokes created about the offspring of interspecific crosses, they are usually based on a biological nonevent.

INTRASPECIFIC VARIABILITY

Over a long course of time people have made the general observation that among sexually reproducing species no two individuals ever look exactly the same. Human beings clearly demonstrate this generalization, which can be verified at any large gathering of people (see Figure 5.1). Although not so readily apparent, the same generalization is valid for other species, also. People who have worked closely with other species usually learn to recognize different individuals in much the same way that we have learned to recognize individual people.

Identical twins provide an exception to the generalization, because these two individuals do look exactly alike, but an understand-

Figure 5.1

No two individuals are ever exactly alike. A quick survey of these nine people reveals the many ways in which individual characteristics may vary. [Photos courtesy of Matthew E. Stockard.]

ing of the information in the preceding chapter should explain why this is so. Identical twins start out as a single zygote, which after several mitotic replications separates into two (or more) distinct developing entities. Each one continues to undergo repeated mitotic replications with development. Because both individuals have identical nuclear controlling mechanisms, and all subsequent growth is produced by mitosis, identical twins are genetically only one individual.

Despite the great diversity that exists among individuals, it has also been observed that there is a strong tendency for closely related individuals to resemble one another to a much greater degree than do nonrelatives (see Figure 5.2). Offspring are much more likely to look like their parents than to look like some total stranger. This is such a common observation in humans that it has been incorporated into a number of colloquial maxims. "She has her grandmother's nose"; "Blood is thicker than water"; "Like father, like son" are all classic examples.

While individual offspring show a high degree of similarity to their parents, they are not identical to either of them, nor are they identical to any other offspring of the same parents. The initial observation of individual variability still holds true, but there are degrees of differ-

Figure 5.2

(a) (b) (c)

Distinctive traits can often be traced through several generations of a family. (a) Charles V, Holy Roman emperor in the sixteenth century, had a distinctive jawline and protuberant lower lip; (b) his son, Philip II, king of Spain, showed the same traits; (c) the same features can also be seen in Philip's great-grandson, Charles II, a later king of Spain. [The Granger Collection, New York.]

ences, and close relatives tend to be less different than nonrelatives. The implication of this pair of observations is that there are some controls on the appearance of offspring that are biologically imposed by something determining the appearance of the parents. An understanding of the regulation of diversity within a species should aid our comprehension of the reproductive process and add to the concept of the continuity of life.

Most of us would like to come to grips with the laws of inheritance from a purely human perspective in order to determine why some people have blonde hair or black hair, blue eyes or brown eyes, or why we have noses, ears, or faces of one shape as opposed to some other shape. For some very obvious, practical reasons, human beings are unsatisfactory subjects to study in determining the basic laws of heredity. For one thing, they take entirely too long to reproduce. An individual has to be 15 or 20 years old before it is capable of reproducing, and then it takes another 15 to 20 years before the next generation is capable of reproducing. That time period does not lend itself to the rapid accumulation of basic observations of fact. A second point is that a specific cross usually results in the production of a limited number of offspring. Humans seldom give birth to more than one child at a time, and it requires nearly a year to produce each one. Because we already know that the offspring exhibit diversity, a sample of one is obviously insufficient. Finally, it is difficult to arrange crosses between individuals on strictly genetic considerations. Cultural restrictions, for example, preclude making crosses between offspring and parent or among offspring of the same parents.

Mendelian Genetics

GREGOR MENDEL AND GARDEN PEAS

The discovery of the basic laws of inheritance can be credited to an obscure Augustinian monk named Gregor Mendel (1822–1884). When not teaching physics and mathematics to schoolchildren and performing other normal clerical activities, Brother Gregor experimented with the controlled crossbreeding of peas in the monastery garden. In 1865 he presented the results of his experiments at a meeting of the scientific society of Bruun (Moravia), which were published in the following year. As near as can be determined at this point, the paper was universally ignored until three separate investigators independently made the same discoveries 35 years later. Although Mendel had died 16 years earlier, all three workers were quick to acknowledge the priority of Mendel's work.

During Mendel's time there were many **true-breeding** strains of garden peas, just as there are today. The strains differed in a number of obvious characteristics such as flower color, height of plant, pea color and shape, pod color and shape, and flower location, but each of these characteristics was expressed consistently in each generation within each strain of garden pea. Some strains of peas consistently produced tall, viney plants, while other strains always produced short, bushy plants. Some strains produced red flowers while others produced white flowers. For each of the characteristics mentioned there existed two alternative forms of expression.

Mendel knew that garden peas were ordinarily **self-pollinating**, which is to say that the plants fertilize themselves to produce the next generation. That is, the male gametes (pollen) produced by a flower usually fertilize the female gametes (egg) produced by the same flower. The resulting zygote eventually grows to an embryo within a seed, which in this case is called a pea. When a red-flowered plant pollinates itself, all the offspring bear red flowers, and when a white-flowered plant pollinates itself, all its offspring bear white flowers. This latter phenomenon is called true breeding.

MONOHYBRID CROSS

What would happen if a red-flowered plant were used to pollinate a white-flowered plant, and vice versa? Or, similarly, what would the offspring be like if a tall plant were used to pollinate a short plant, and vice versa? Mendel carried out several crosses of this type involving contrasting expressions of a given characteristic, which is called a **monohybrid cross.** In every case he observed that the offspring showed only one of the two alternative expressions of the characteristic being studied. For example, when the parental generation (or P_1) consisted of a true-breeding, red-flowered plant and a true-breeding, white-flowered plant, the first generation of offspring (called the first filial generation, or F_1) consisted of all individual pea plants bearing red flowers (see Figure 5.3). Similarly, when the P_1 consisted of tall plants and short plants, the F_1 consisted of individual plants that were all tall. Repeatedly, when Mendel performed monohybrid crosses between parents with contrasting expressions of the same characteristic, only one of those two expressions appeared in the first filial generation. It made no difference which parent contributed the pollen and which received the pollen; the same expression of the characteristic showed up in the offspring.

Dominance One of the two expressions of each characteristic seemed to take precedence over the other. For example, red flower color

Figure 5.3

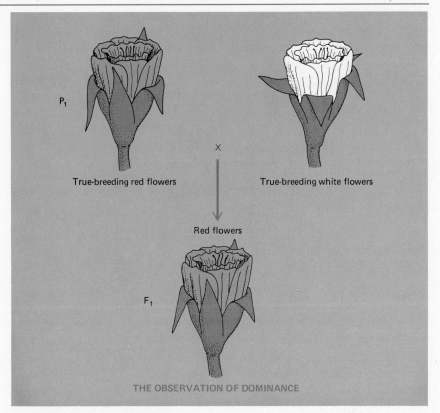

The observation of dominance.

seemed to take precedence over white flower color. This can be expressed in a slightly different way by saying that the color red in pea flowers seems to be **dominant** over the color white in pea flowers. Similarly, tallness in pea plants appears to be dominant over shortness in pea plants. One can also look at it the other way around and consider white flower color or short plant height as **recessive** to red flower color or tall plant height (see Figure 5.4).

When plants bearing yellow pods were crossed with plants bearing green pods, the offspring all grew up into plants that bore green pods. Green pod color is said to be dominant over yellow pod color. By making a number of such crosses, Mendel discovered that the round form of peas is dominant to the wrinkled form of peas, inflated pod shape is dominant to constricted pod shape, yellow peas are dominant

Figure 5.4

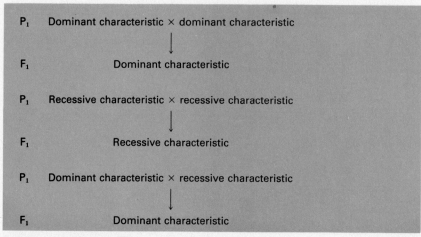

Mendel's observation.

to green peas, and axillary flowers are dominant to terminally located flowers. Notice in particular that dominance is a relative term for each particular trait. While green is dominant to yellow in pod color, it is recessive to yellow in pea color.

Phenotype   The outward expression of a particular characteristic is called its **phenotype.** A phenotype is simply the way the character appears. We can describe all Mendel's F_1 offspring as demonstrating the dominant phenotype for each of the characters studied. When you cross a true-breeding, dominant phenotype with a true-breeding, recessive phenotype, you always get a dominant phenotype in the offspring.

Having determined that in each case the first filial generation resembles only the dominant parental phenotype, Mendel then proceeded to allow the F_1 to fertilize themselves normally. The F_1 thus served as the parents for a second filial generation, which can be called the F_2. In examining this second filial generation, Mendel consistently found that both phenotypes of each of the characteristics appeared. Plants showing the dominant phenotype and plants showing the recessive phenotype appeared in the F_2 in the proportion of three dominants to one recessive (see Figure 5.5).

Let us look at a specific example. Among 929 plants in an F_2, 705 bore red flowers while 224 had white flowers. In a separate experiment involving 1064 plants in the F_2, 787 showed the dominant tall pheno-

Figure 5.5

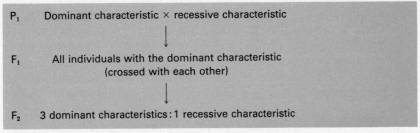

P_1 Dominant characteristic × recessive characteristic

F_1 All individuals with the dominant characteristic
            (crossed with each other)

F_2 3 dominant characteristics : 1 recessive characteristic

A generalized phenotypic description of any monohybrid cross.

type while 277 showed the recessive short phenotype. What is important to notice here is that phenotypes that are not found in the F_1 do appear in the F_2.

Mendel then allowed each plant in the F_2 to pollinate itself, providing an F_3. All plants in the F_2 that showed the recessive phenotype produced offspring in the F_3 that all had the same recessive phenotype. Of the plants in the second filial generation that showed the dominant phenotype, about a third produced an F_3 in which all the plants expressed the dominant phenotype, while the remaining two-thirds produced offspring that expressed a mixture of dominant and recessive phenotypes in a ratio of 3 : 1.

The mathematical regularity of these results, regardless of which specific character is being considered, implies that some one general process is at work. The basic outlines of the process must conform to the following conclusions, derived from the experimental observations:

1. Each pea plant must have two hereditary factors for each character in question, such as flower color or plant height.
2. Hereditary factors can exist in more than one form, which express themselves in different ways.
3. When gametes are formed, the two hereditary factors separate into different gametes such that each gamete has only one hereditary factor for each character.
4. In the production of a new individual in the next generation formed by the fusion of gametes, each new plant receives one hereditary factor from the gamete of each parent, thus restoring the existence of two hereditary factors for each character in each plant.
5. Since one of the two contrasting phenotypes of the parental generation can disappear in the first filial generation and reappear in the second filial generation, then the hereditary factors must exist as particulate structures in the plant.

Genes   Subsequent workers in the area of genetics have replaced Mendel's expression *hereditary factors* with the word **gene.** A gene is a unit of inheritance that in the present context can be thought of as being responsible for the production of a character such as flower color, pod shape, or plant height. The gene, based upon observed breeding data, is an abstract entity whose existence is inferred as the simplest explanation of the results. A gene may exist in alternative forms, each of which is called an **allele.** In the case of the flower color gene in garden peas, there is a dominant red allele and a recessive white allele.

Genes can be represented symbolically by using letters (or combinations of letters) of the alphabet. Such a symbolic representation of the presumed genetic makeup of an organism is called the **genotype.** Conventionally capital letters are used to signify the dominant alleles, while lowercase letters are used to signify the recessive alleles. If an organism has two genes for a particular character, and both genes exist in the same allelic state (for example, *AA* or *aa*), then the genotype is said to be **homozygous** or homozygotic. If both genes occur in different allelic states (for example, *Aa* or *aA*), then the genotype is called **heterozygous** or heterozygotic.

It is now possible to repeat Mendel's crosses in abstract, symbolic terms (see Figure 5.6). An individual that is homozygous dominant for the trait tall plant (genotype *HH*) is crossed with an individual that is homozygous recessive for the trait short plant (genotype *hh*). The homozygous dominant parent produces gametes all of which carry the *H* allele. The homozygous recessive parent produces gametes all of which carry the *h* allele. Fusion of the gametes from each parent produces a new generation of heterozygous offspring (genotype *Hh*). Because the *H* allele is dominant to the *h* allele, the offspring are phenotypically dominant. These heterozygous F_1 pea plants now produce both male and female gametes, which will combine to produce the second filial generation. Each gamete produced must contain a gene responsible for plant height, but it can be of either allelic form. For every gamete containing an *H* allele, there must be another containing an *h* allele. Equal numbers of *H*-bearing and *h*-bearing male gametes must interact with a population of female gametes composed of equal numbers of those bearing *h* or *H*. If fertilization occurs at random with respect to the genotypes of the gametes, then there should be an equal production of *HH*, *Hh, hH,* and *hh* genotypes in the second filial generation. The homozygous dominant and two heterozygous genotypes will all be phenotypically tall, while the homozygous recessive genotype will be phenotypically short. This accounts for Mendel's observed phenotypic ratio of 3:1. If the homozygous recessive (phenotypically short) individuals are used to produce a third filial generation, all the gametes involved will

Figure 5.6

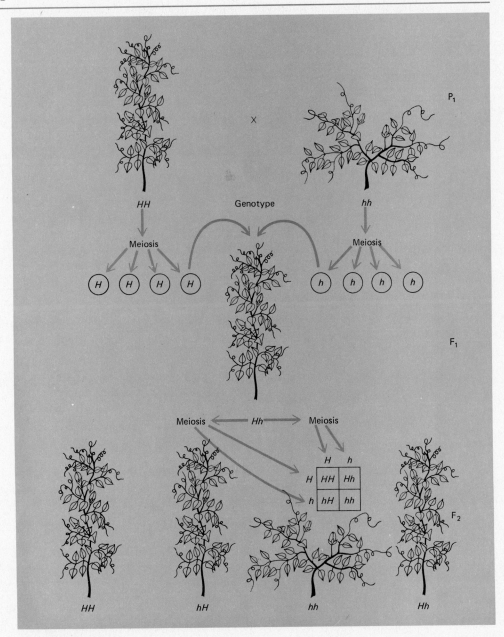

Monohybrid cross.

carry the recessive allele for plant height (h), meaning that all the offspring will have the genotype hh and will be phenotypically short. If
the homozygous dominant (one-third of the phenotypically tall) individuals are used to produce a third filial generation, all the gametes involved will carry the dominant allele for plant height (H), meaning that
all their offspring will have the genotype HH and will be phenotypically tall. If the heterozygous individuals (the remaining two-thirds of
the dominant phenotypes in the F_2) are used to produce an F_3, the results should be exactly the same as those for the crossing of heterozygotes to obtain the F_2; namely, a ratio of three dominant phenotypes to
one recessive phenotype.

Ratios and Probabilities   It is important to pause for a moment at this
point to emphasize the idea that the predictability of phenotypic results is probabilistic. We are assuming that the combining of gametes to
form a new generation of offspring is a completely chance event controlled only by the relative abundance of the gametic genotypes. If one-
half of the male gametes are of a given genotype and one-half of the
female gametes are of the same genotype, then we expect one-fourth of
the resulting paired combinations to consist of that particular kind of
pairing. It is a situation analogous to the tossing of coins.

Imagine that you have two coins, a nickel and a penny, each of
which has a head and a tail. If you were to flip the penny, half of the
time it would land head up and the other half of the time it would land
tail up. The same thing would be true for the nickel. The probability of
tossing either coin and coming up with a head is one-half. Suppose you
flipped your penny and it landed with the head up. What effect would
this have on the results of flipping your nickel? None whatsoever. The
nickel still has a probability of one-half of landing head up. Thus the
probability of tossing both coins and having both land heads up is
$\frac{1}{2} \times \frac{1}{2}$, or $\frac{1}{4}$. The same probability exists for having both coins land with
their tails up. In fact, the same probability exists for any specific combi-
nation of results (see Figure 5.7).

Suppose now that we flip our two coins four times. Does this
mean that we will get one head-head combination, one head-tail combination, one tail-head combination, and one tail-tail combination? Not
necessarily. Each of the paired tossings is an independent event and
thus has no influence on the outcome of any other independent event.
Coming up with a head-head combination does not prevent you in any
way from coming up with another head-head combination. All the laws
of probability say is that with an infinite number of paired tosses you
can expect to come up with head-head combinations $\frac{1}{4}$ of the time.

In the second filial generation of a monohybrid cross, we expect to

Figure 5.7

	COIN A	
	Probability of flipping a head = 0.5	Probability of flipping a tail = 0.5
COIN B Probability of flipping a head = 0.5	Probability of flipping a head on A and a head on B = 0.25	Probability of flipping a tail on A and a head on B = 0.25
Probability of flipping a tail = 0.5	Probability of flipping a head on A and a tail on B = 0.25	Probability of flipping a tail on A and a tail on B = 0.25

Probability in flipping coins.

find that $\frac{1}{4}$ of the offspring are homozygous dominant, $\frac{1}{4}$ are homozygous recessive, and $\frac{1}{2}$ are heterozygous. The genotypes Hh and hH are genetically indistinguishable, so the probability of finding heterozygous offspring becomes $\frac{1}{4}$ plus $\frac{1}{4}$. The homozygous dominant genotype is phenotypically indistinguishable from the heterozygous genotype, so we would expect to find $\frac{3}{4}$ ($\frac{1}{4}$ plus $\frac{1}{2}$) of the offspring to be phenotypically dominant. This is, of course, assuming that an infinite number of trials are attempted. As most organisms are not capable of producing an infinite number of offspring, slight deviations from the expectations may be found. As a general rule of thumb, the smaller the sample size, the greater is the chance of deviation from the expected results. Obviously it is not possible to obtain a three-to-one ratio in the F_2 if there are only two offspring.

Intermediate Inheritance   While the ideas of dominance and recessiveness are very common in discussing the inheritance of genes, there have been several examples discovered subsequent to the work of Mendel in which dominance and recessiveness did not exist. For example, in horses and cattle there is a simple Mendelian gene that controls coat color. One of the alleles is responsible for a coat color that can be called red, while its alternative is responsible for white coat color. If one crosses a red male with a white female (or vice versa), the offspring all have a coat color that is a mixture of red and white hairs, a phenotype called roan. Crossing roan males with roan females results in the production of offspring that are red or roan or white in a $1:2:1$ ratio. The easiest way of interpreting these results is to assume that red individuals are homozygous for an allele that we'll call R. White individuals are homozygous for another allele that we'll call r. Offspring produced from such a mating will all be heterozygous with a phenotype that can be referred to as intermediate. When the heterozygous individuals are allowed to interbreed, $\frac{1}{4}$ of their offspring are homozygous for the R allele, $\frac{1}{4}$ of the offspring are homozygous for the r allele, and $\frac{1}{2}$ of the offspring are heterozygous, as are their parents. This situation in which neither allele expresses dominance over the other can be referred to variously as intermediate inheritance, incomplete dominance, or codominance. The name you give to this phenomenon is less important than recognizing that the heterozygous genotypes express a phenotype that is different from, and in between that of, the two homozygous genotypes.

What is particularly interesting about intermediate inheritance is that the phenotypes of the organisms are a direct reflection of their genotypes. Consequently, the phenotypic ratios expressed in a given generation are precisely the same as the genotypic ratios in that generation. With intermediate inheritance it is possible for one gene having two alleles to be responsible for three different phenotypes.

TEST CROSS

In those cases where one allele is dominant to another, the phenotype is not a direct reflection of the genotype, except in the case of the homozygous recessive. If an organism has a dominant phenotype, it is not possible to determine by inspection whether that individual is homozygous or heterozygous for the dominant allele. All that can be determined by examining the phenotype is that the organism has at least one dominant allele.

There are times when it would be very useful to know whether a particular individual with a dominant phenotype is homozygous or heterozygous. The easiest way to make such a determination is to arrange a cross with a recessive phenotype, which is called a **test cross.** An individual with the recessive phenotype must have a homozygous recessive genotype, and thus it is only capable of producing gametes that carry the recessive allele. If the dominant phenotype is due to the homozygous condition, then all the gametes produced by this individual must carry the dominant allele. All the offspring produced must be of the heterozygous genotype and show the dominant phenotype. If, on the other hand, the dominant phenotype is due to the heterozygous condition, then the individual can produce two kinds of gametes in equal proportions: those carrying the dominant allele and those carrying the recessive allele. Offspring produced by fusion of these gametes with those of the recessive parent must also be of two kinds in equal proportions: heterozygous offspring showing the dominant phenotype and homozygous individuals showing the recessive phenotype (see Figure 5.8).

A test cross can give two entirely different kinds of results depending on the genotype of the phenotypically dominant parent. If all the offspring show the dominant phenotype, it is easiest to conclude that the phenotypically dominant parent is genotypically homozygous. If the offspring show an equal mixture of dominant and recessive phenotypes, then it must be concluded that the phenotypically dominant parent is genotypically heterozygous.

DIHYBRID CROSS

Mendel's description of the patterns of inheritance of single genes was highly significant, but it was not his only contribution. He also performed breeding experiments with peas in which be observed the transmission of two characteristics simultaneously. These are called **dihybrid crosses.** In one of his experiments he used, as the parental generation, one true-breeding strain that produced round, yellow peas and another strain that produced wrinkled, green peas. From his monohybrid crosses it was known that round is dominant to wrinkled and yellow is dominant to green. It should thus come as no surprise that the phenotypes of the first filial generation were round and yellow peas. When these plants were used to pollinate themselves, producing a second filial generation, the observed phenotypes were as follows: 315 round and yellow, 101 wrinkled and yellow, 108 round and green, and 32 wrinkled and green. It is particularly interesting to note that crossing within the F_1 produces some brand new phenotypic combinations not encountered in previous generations; namely, wrinkled-yellow and round-green.

These results indicate that the genes for pea color and pea shape are independent; they are inherited as separate entities. The easiest way to understand what has happened in this set of crosses is to make a symbolic approach as in Figure 5.9. One of the original parents consists of plants with the genotype $RRYY$. The R gene is for pea shape and is homozygous for the round allele, while the Y gene stands for pea color and is homozygous for the yellow allele. The other parent is homozygous for the recessive alleles of both of these genes and has the genotype $rryy$. The first parent must produce gametes containing one of

Figure 5.8

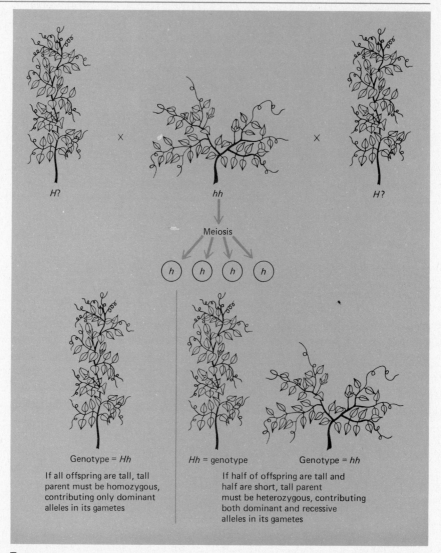

Genotype = *Hh*

If all offspring are tall, tall
parent must be homozygous,
contributing only dominant
alleles in its gametes

Hh = genotype

Genotype = *hh*

If half of offspring are tall and
half are short, tall parent
must be heterozygous, contributing
both dominant and recessive
alleles in its gametes

Test cross.

each kind of gene; hence their genotype must be *RY*. The second parent's gametic genotype must be *ry*. The fusion of these two gametes results in the production of the first filial generation, all individuals of which have the genotype *RrYy*. They are heterozygous for both genes.

Figure 5.9

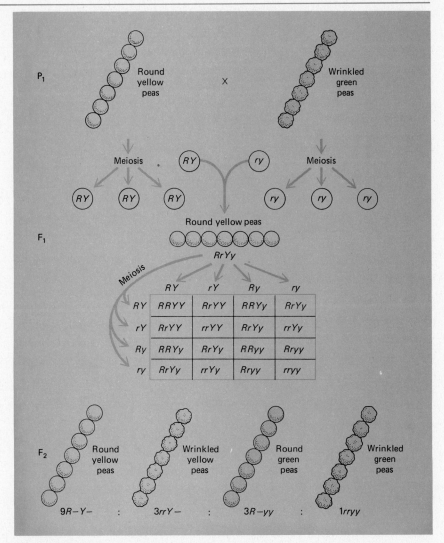

Dihybrid cross.

Because both the alleles for round shape and yellow color are dominant, the plants express the dominant phenotypes for both characters.

Now what happens when the F₁ produces gametes? The F₁ genotype $RrYy$ must produce gametes that have one of each kind of gene. Thus the male gametes can be of 4 different kinds: RY, Ry, rY, or ry.

The same thing is true for the female gametes. Gametes with the genotype Rr or Yy are not possible because they violate the requirement that the gametes must contain one of each kind of gene. If there are 4 different kinds of genotypically distinct gametes within each sex, then these can combine together to form 16 different genotypes in the second filial generation. Nine of the 16 genotypes are phenotypically equivalent in that they all show both characteristics in the dominant condition. Three of the 16 genotypes result in a phenotype that is dominant for one of the characteristics and recessive for the other, while another 3 of the 16 genotypes result in a phenotype that is dominant for the other characteristic and recessive for the first. The remaining genotype of the 16 is homozygous for both kinds of recessive alleles and thus has the phenotype of both recessive characteristics.

We would expect the F_2 of a dihybrid cross to have $\frac{9}{16}$ of its makeup consisting of individuals showing both dominant traits, $\frac{3}{16}$ of the individuals showing one of the dominant traits, $\frac{3}{16}$ of the individuals showing the other dominant trait, and $\frac{1}{16}$ of the individuals showing no dominant traits. In other words, we would expect a phenotypic ratio of 9:3:3:1. An examination of Mendel's actual results shows a very close approximation to these expectations. (The observed ratio is 9.06:2.91:3.11:0.92). This close a fit to the expected results indicates that the probabilistic explanation of the behavior of genes is most likely correct.

MULTIPLE-TRAIT CROSSES

If our earlier conclusions about the behavior of genes is correct, we should be able to predict the outcome of a cross involving three (or more) genes simultaneously. Suppose we were to arrange a cross between garden pea plants that produce round, yellow peas and red flowers, and plants that produce wrinkled, green peas and white flowers. We would expect the offspring to be heterozygous for all three genes and to be phenotypically dominant for all three traits. That is, all the plants in the F_1 would produce round, yellow peas and bear red flowers. The second filial generation would be produced by the fusion of gametes from the F_1, of which there should be 8 genotypically different kinds: RYF, RYf, RyF, Ryf, rYF, rYf, ryF, and ryf. The random fusion of these kinds of gametes can result in 64 different combinations of genotypes, which appear as eight different phenotypes in a ratio of 27:9:9:9:3:3:3:1 (see Figure 5.10).

The number of different kinds of gametes that can be produced is a function of the number of different genes that are in the heterozygous state. Each heterozygous genotype can produce two kinds of gametes. Thus the number of gametic genotypes that can be produced will equal

Figure 5.10

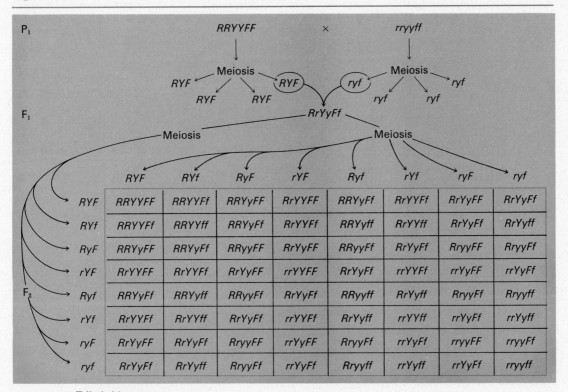

Trihybrid cross.

2^n, where n equals the number of genes in the heterozygous state. If the gametes combine at random, they will produce $2^n \times 2^n$ different combinations of genotypes. The number of kinds of phenotypes and their relative proportions will depend upon the presence or absence of dominance, but if we assume dominance does exist for all the genes observed, then the phenotypic proportions should conform to the laws of the **binomial expansion.** If A stands for the dominant allele of any gene, a stands for the recessive allele of any gene, and n stands for the number of genes being observed, then $(3A + a)^n$ will represent the proportion of different phenotypes in the F_2.

Mendel actually performed a **trihybrid** cross and observed the following phenotypic results: 269 round, yellow peas with red flowers; 98 round, yellow peas with white flowers; 86 round, green peas with red flowers; 88 wrinkled, yellow peas with red flowers; 27 round, green peas with white flowers; 34 wrinkled, yellow peas with white flowers;

30 wrinkled, green peas with red flowers; and 7 wrinkled, green peas with white flowers. These results conform very closely to the expected ratio of $27:9:9:9:3:3:3:1$ (see Figure 5.10).

Rediscovery and Refinement of Mendelism

GENE-CHROMOSOME ASSOCIATION

With the rediscovery of Mendel's work in the early twentieth century, it became rapidly apparent that the behavior of the abstract genes was precisely duplicated by the observed behavior of the cellular constituents mentioned in Chapter 4, the chromosomes. In higher organisms, such as pea plants, there are two of each kind of chromosome in each cell, in much the same way that Mendel inferred that each plant must have two of each kind of gene. In the process of meiosis each resulting haploid cell (gamete) has only one of each kind of chromosome, which is precisely what Mendel concluded must happen to genes when gametes are formed. The diploid number of chromosomes (two of each kind) is restored by fertilization, which Mendel concluded must happen for genes in the formation of a new generation of pea plants. The independent assortment of the seven characteristics studied by Mendel, coupled with the observation that garden peas have seven different kinds of chromosomes, led to the reasonable inference that each of Mendel's seven proposed genes must be associated in some way with a different kind of chromosome. The first temptation is to guess that genes and chromosomes are the same thing.

While the early geneticists quickly established a relationship between genes and chromosomes, two lines of evidence indicated that they were not exactly the same thing. The first line of evidence was that for most organisms studied extensively the number of genes was much greater than the number of kinds of chromosomes. For example, in the well-studied fruit fly (*Drosophila melanogaster*), several hundred different genes are known to exist, although there are only four different kinds of chromosomes. The second line of evidence is in the form of unexpected phenotypic ratios in the F_2 of certain kinds of crosses.

AUTOSOMAL LINKAGE AND GENE MAPPING

In sweet peas it is established that purple flower color is dominant to red flower color and that the production of oval (or long) pollen grains is dominant to the production of round pollen grains. If you were to cross sweet peas having purple flowers that produce long pollen grains with sweet peas that have red flowers and produce round pollen grains and then were to observe the offspring in the *second* filial generation, you will probably not observe the expected $9:3:3:1$ ratio. Instead, the offspring usually appear in a phenotypic ratio of $30:3:3:8$. Such results are totally unexpected and not predictable. How can they be explained?

Instead of allowing the first filial generation, which is heterozygous for both genes, to breed with each other, thus producing a second filial generation, let us cross the F_1 to the recessive parental phenotype. This would constitute a dihybrid test cross. The advantage of such a cross is that the recessive parental phenotype can only produce one kind of gamete; those possessing the recessive alleles for flower color and pollen shape. If there are two independent genes controlling flower color and pollen shape, then meiosis should produce four genotypically distinct kinds of gametes from the F_1, which will combine with the one kind of gamete from the recessive parental type to produce four phenotypically different kinds of offspring in a ratio of $1:1:1:1$ (see Figure 5.11). However, if there is only one gene controlling these two traits, then meiosis in the F_1 can only produce two kinds of genotypically distinct gametes, which will combine with those of the recessive parental type to produce two phenotypically different offspring in the ratio of $1:1$ (see Figure 5.12). The observed results conform to neither set of expectations. The offspring from such a cross occur in a phenotypic ratio of 7 purple-flowered, long pollen:1 purple-flowered, round pollen:1 red-flowered, long pollen:7 red-flowered, round pollen. Most of the offspring display the two **parental phenotypes,** indicat-

Figure 5.11

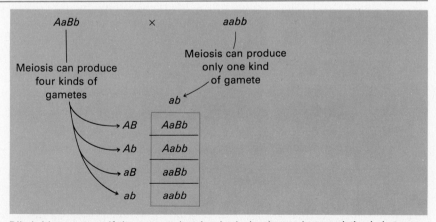

Dihybrid test cross. If the parent showing both dominant characteristics is heterozygous for both genes, then fertilization produces four phenotypically distinct kinds of offspring. How many kinds of offspring would result if the dominant parent were heterozygous for only one gene? For none?

Figure 5.12

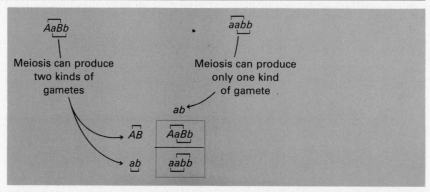

Dihybrid test cross with linked genes. If the parent showing both dominant characteristics is heterozygous for both linked genes, then fertilization produces two phenotypically distinct kinds of offspring, which both look like the parents.

ing that the two traits are not independent. But a small proportion of the offspring display the **recombinant phenotypes,** arguing for the existence of two different genes. Two out of every 16 offspring (12.5 percent of the offspring) evince the existence of two independent genes, while 87.5 percent of the offspring indicate that only one gene is involved. If there were two genes, but both were located on the same chromosome, then we would expect them to segregate during meiosis as if they were only one gene.

To understand these strange ratios, we must review the process of meiosis. During prophase I, while the homologous chromosomes are in synapsis, the various parts of the two chromosomes get very wrapped up in one another. Part of one chromosome frequently overlaps with a similar part of its homologue, forming what is called a **crossover** or a **chiasma** (see Figure 5.13). It often happens that these entanglements are so severe that when the homologous chromosomes separate during anaphase I, the chromosomes break at the crossover and exchange ends. This means that a part of the maternal chromosome is attached to the paternal chromosome, and vice versa.

If genes are arranged as a linear sequence on the arms of chromosomes, the occurrence of crossovers between the sites of their locations could account for the recombinant phenotypes in the offspring of the dihybrid test cross. If the genes for flower color and pollen shape in sweet peas are located on the same chromosome, then the dominant allele for flower color is physically united to the dominant allele for pollen shape. The genes are said to be linked. A gamete formed that con-

Figure 5.13

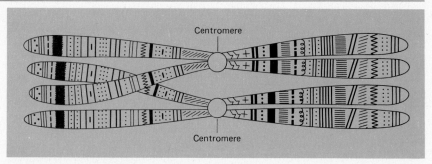

Centromere

Centromere

A chiasma or crossover.

tains the dominant allele for flower color must also contain the dominant allele for pollen shape. Any gamete containing the recessive allele for one of those two genes must also contain the recessive allele for the other. When a crossover occurs in an organism heterozygous for two linked genes, and the crossover occurs between the location of those two linked genes, the chromosomes that result carry the recessive allele for one gene and the dominant allele of the other gene, and vice versa. Every time a crossover occurs in such a heterozygous individual, two recombinant gametes are formed (see Figure 5.14).

Crossovers occur during meiosis with a very high level of frequency. They can be found in almost any cell undergoing meiosis. However, any crossover occurring anywhere but along the distance on the chromosome between the two particular genes that are being studied will not affect the outcome of the dihybrid test cross. Clearly the chief factor that determines the frequency with which a crossover will occur between two particular genes on a given chromosome is the distance between those two genes. If the distance between the two genes is very small, the chances of a crossover occurring in that specific distance are extremely limited, whereas the chances of a crossover occurring between the two genes are very great if the distance between the genes is large. This fact means it is possible to use the frequency of crossovers as an indirect measure of the distance between two genes on a chromosome.

Using the genes for flower color and pollen shape in sweet peas as an example, let me demonstrate how to calculate the distance separating linked genes. In the cross *PpLl* × *ppll* we obtained offspring in the ratio 7*PpLl* : 1*Ppll* : 1*ppLl* : 7*ppll*. Only the purple-round (*Ppll*) and red-long (*ppLl*) offspring demonstrate the recombinant phenotypes. As

Figure 5.14

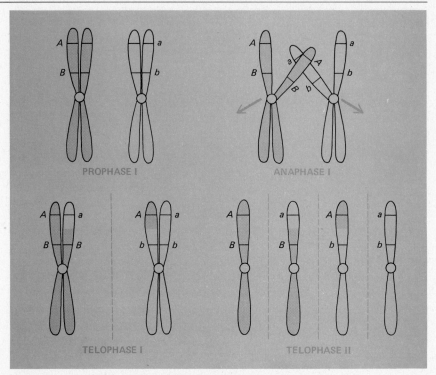

Formation of recombinant gametes from linked genes.

these constitute $\frac{2}{16}$, or $12\frac{1}{2}$ percent, of the offspring, we are forced to conclude that recombination between these two genes has occurred in only $12\frac{1}{2}$ percent of the meiotic processes. If the two genes are linked, the only event that can result in recombination during meiosis is crossing-over. This means that crossovers must have occurred between these two genes in $12\frac{1}{2}$ percent of the meiotic events. Because the frequency of crossing-over is directly dependent upon the distance between the two genes, it is possible to create an arbitrary unit of measurement, called the **map unit,** and conclude that the genes for flower color and pollen shape in sweet peas are separated by a distance of $12\frac{1}{2}$ map units. Nothing in this train of reasoning allows us to convert map units into traditional units of measurement such as the metric or English systems.

If one studies enough different genes in the same organism, it will become evident that groups of genes are linked together. The number of these linkage groups will be equal to the number of kinds of chromosomes present in that species of organism. Dihybrid test crosses can be used to measure the distances between each of the genes, and these distances can be used to determine the relative positions of the various genes along the length of the chromosome; a process called chromosomal mapping. Let me illustrate with a purely imaginary example.

Suppose we have the following results from a series of crosses:

AaBb × *aabb* yields	1:1:1:1	independent segregation
AaCc × *aacc* yields	18:7:7:18	28% recombinants
AaDd × *aadd* yields	4:3:3:4	42% recombinants
AaEe × *aaee* yields	49:1:1:49	2% recombinants
BbCc × *bbcc* yields	3:1:1:3	25% recombinants
BbDd × *bbdd* yields	9:1:1:9	10% recombinants
BbEe × *bbee* yields	1:1:1:1	independent segregation
CcDd × *ccdd* yields	6:1:1:6	14% recombinants
CcEe × *ccee* yields	3:1:1:3	25% recombinants
DdEe × *ddee* yields	3:2:2:3	40% recombinants

A first inspection would lead you to suspect that the *A* and *E* genes are not linked to the *B* gene because they assort independently. However, further inspection reveals that the *A* and *E* genes are both linked to the *C* and *D* genes, which are, in turn, linked to the *B* gene. *A* and *C* are separated by 28 map units and *A* and *D* are separated by 42 map units. Thus the sequence of genes is either *ACD* or *DAC*. Because genes *C* and *D* are separated by 14 map units, the sequence *ACD* is much more probable. In similar fashion, one can determine that the relative positions of the other genes form a sequence of *AECDB*. Notice, in particular, that when linked genes are 50 map units or more apart, they result in a phenotypic test cross ratio of 1:1:1:1. Thus 50 map units of distance on a chromosome is enough to ensure that crossovers will occur half the time.

SEX DETERMINATION

It was noted earlier that chromosomes occur in pairs in all diploid cells. Each member of the pair looks alike, and there are two of each kind of chromosome. While that is generally true for most kinds of chromosomes in most kinds of organisms, there are some notable exceptions. In some animals where the sexes are different—that is, in those cases where there are distinctly different individuals that are either male or female—there is one sex in which there are unpaired chromosomes, or at least two chromosomes that are not alike in shape and size. Such chromosomes are called **sex chromosomes.** All chromosomes that are not sex chromosomes are called autosomal chromosomes or **autosomes** (see Figure 5.15).

In many cases, including human beings, the sex chromosomes are

Figure 5.15

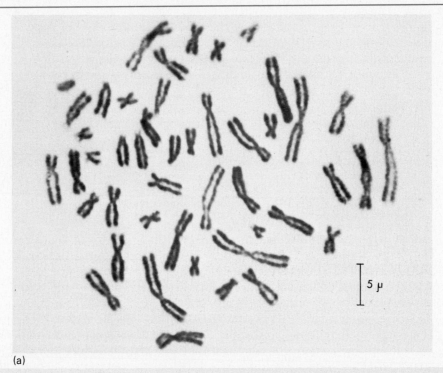

(a)

(b)

(a) The chromosomes of a normal human male. There are 46 chromosomes visible. (b) The chromosomes are sorted and arranged by size and centromere position. There are 22 pairs of autosomal chromosomes, one X chromosome (in group C), and one Y chromosome (in group G). [Photo courtesy of Carolina Biological Supply Company.]

Figure 5.16

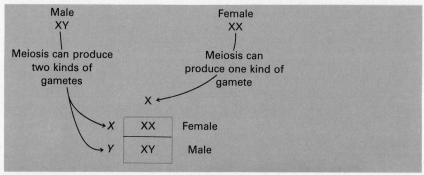

Sex determination.

called X and Y. Each cell in an adult human being has 46 chromosomes; 22 pairs of autosomes and 2 sex chromosomes. Females have 2 X chromosomes. They appear and behave just as though they were standard autosomes. In males, however, the sex chromosomes consist of one X and one Y. It appears as though this chromosomal configuration is responsible for sex determination. When females undergo meiosis, they are capable of producing only one kind of gamete with respect to the sex chromosome contained. All female gametes contain one X chromosome and 22 autosomes. However, when males undergo meiosis, the X and Y chromosomes form a synapsis resulting in gametes that contain either an X or a Y chromosome with 22 different autosomes. Fertilization of a female gamete by an X-bearing male gamete results in the establishment of a zygote with 46 chromosomes, including 2 X chromosomes (female). The fertilization of an egg by a Y-bearing male gamete results in a zygote with 46 chromosomes, including an X and a Y, which is male (see Figure 5.16).

SEX-LINKED GENES

It has been known for some time that genes can be associated with the sex chromosomes as well as with the autosomes. Such genes show a pattern of inheritance that is strikingly different from that observed in genes associated with autosomal chromosomes. A classic example of such a sex-linked gene involves white eye color in the fruit fly *Drosophila melanogaster*. If you cross red-eyed female flies with white-eyed male flies, all the offspring have red eyes. We conclude from this that the red-eyed allele is dominant to the white-eyed allele. If we now cross the males of this F_1 with the females of the F_1 (their sisters), we observe that the offspring show a phenotypic ratio of three red-eyed flies to one white-eyed fly. This is just what you should expect to find in a monohybrid cross, except that all the white-eyed flies are male. Careful inspection reveals that the ratio is 2 red-eyed females:1 red-eyed male:1 white-eyed male.

The **reciprocal cross** gives very different results. When white-eyed female flies are crossed with red-eyed male flies, the first filial generation consists of half red-eyed flies and half white-eyed flies. All the females of the F_1 have red eyes and all the males have white eyes. If these flies are used to produce an F_2, the offspring still show a one-to-one phenotypic ratio with respect to eye color, but there is no longer a correlation with the sex of the flies. In a sense the F_2 of this particular cross show a 1:1:1:1 ratio of red-eyed males:white-eyed males:red-eyed females:white-eyed females.

This observation is totally different from anything experienced by Mendel, in that none of his genetic characteristics were in any way influenced by the sex of the parent. In this case the phenotype of the mother seems to have a very strong effect on the male offspring. The easiest explanation of these results is to assume that the gene for eye color in fruit flies is found on the X chromosome. If we make this as-

Figure 5.17

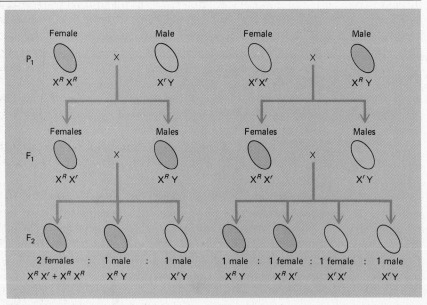

Sex linkage.

sumption, then it is obvious that male flies have only one gene for eye color, and the pattern of inheritance occurs as shown in Figure 5.17. While the inheritance of eye color in fruit flies may not strike you as particularly relevant, the human traits of red-green color blindness, muscular dystrophy, and **hemophilia** (bleeder's disease) are inherited in exactly the same fashion.

GENE INTERACTIONS

The impression that I have tried to give you to this point is that there are a lot of things on chromosomes called genes, which in some way or another determine what the phenotype of the organism is going to be. While in some cases this seems to be a direct one-to-one relationship, there are many other cases where it is more involved.

Gene-Gene Interactions   There are a number of cases known where it takes the cooperative interaction of several genes to produce a given phenotype. Perhaps one of the better examples comes from the study of **comb** shape in chickens. Combs are the fleshy protuberances found on the top of chicken heads, which come in different shapes that are partly characteristic of different breeds of fowl. If one crosses a Wyandotte chicken with a comb shape called rose with a Brahma chicken having a comb shape described as pea, all the offspring will grow up with a comb shape that is called walnut. This comb shape is characteristic of a breed of chicken called Malay. At first glance this situation looks as though it is a case of intermediate inheritance, because two different phenotypes have produced a third phenotype. However, when fowl with walnut combs are allowed to interbreed, their offspring display four different phenotypes in the ratio of 9 walnut : 3 rose : 3 pea : 1 single. Single comb is a shape that is characteristic of Leghorn chickens. The fact that we have a 9:3:3:1 ratio argues strongly that this is not a case of intermediate inheritance but is a dihybrid cross, where both genes affect the same character of comb shape (see Figure 5.18). There must be one gene whose dominant allele produces rose comb while the recessive allele produces single comb. There must be a second gene whose dominant allele produces pea comb while its recessive allele also produces a single comb. When both genes are present in the dominant allelic state, the phenotype expressed is walnut comb. The phenotype of single comb can only be expressed if both genes are present in the homozygous recessive condition.

An even more obvious example of many genes working together to affect the same phenotype can be found in those characters that appear to vary continuously yet still have a strong hereditary basis. Size in human beings appears to be one such characteristic. Large parents

Figure 5.18

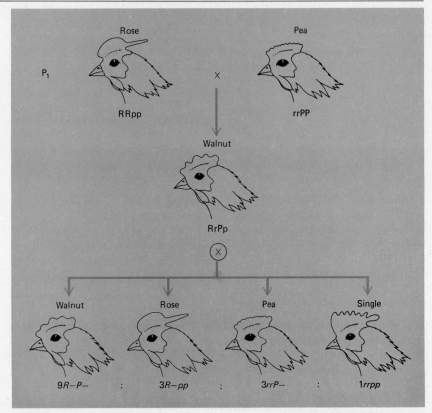

Inheritance of comb shape in fowl.

tend to produce large offspring and small parents tend to produce small offspring, but there are all kinds of exceptions and intermediate conditions. Size in human beings is quite different from the situation previously described in garden peas, where an individual was either tall or short. Not only is there an intermediate phenotype in people, but there are many different intermediate phenotypes. It would appear that the most likely explanation for these multiple phenotypes is some large number of genes that lack dominance between alleles and whose effect on the phenotype is additive. A person homozygous for the tall alleles of 12 different genes for size would be very tall, while a person homozygous for the short alleles of those same genes would be very small. Anyone heterozygous for all the genes, or homozygous tall for half the genes and homozygous short for the other half of the genes, would display the same intermediate phenotype. The random recombination of the various alleles of a large number of genes could easily account for the large amount of variable phenotypic expressions that we find in this character.

Skin color in human beings seems to be determined in a very similar fashion, by the interaction of several different genes. Each gene has one allele controlling the addition of some amount of pigment in the skin, while its alternative allele adds little or no pigment. The additive effect of all the alleles of the different genes is responsible for the intensity of skin color.

Gene-Environment Interactions   To make matters even more complex, it is well known that the working of genes—the conversion of genotype to phenotype—can be strongly influenced by totally nongenetic factors of the environment. Adequate nutrition is obviously an important factor influencing the phenotypic expression of size, and exposure to sunlight can modify the phenotypic expression of skin color.

A more striking example is found in the coat color of domestic rabbits. There is an allele for one gene that, when found in the homozygous condition, results in a coat color that is called himalayan. The himalayan phenotype consists of body fur that is white while the ears, feet, tail, and nose are black. The Himalayan allele is temperature-sensitive, meaning that the black pigment can only be deposited when the surroundings are cooler than a certain temperature, about 33°C (92°F). The extremities of the rabbit are usually below this temperature, so the fur there is black. But the body's temperature is usually higher than 33°, so no pigment is deposited and the fur is white. If the fur is plucked from the rabbit's extremities and allowed to regrow in an environment where the temperature is 35°C or higher, the new fur on the extremities will be white. Similarly, if the fur is plucked from a rabbit's body and kept below 33° through the use of an ice pack, the regenerated hair will be black.

The importance of all these various kinds of complexities is to illustrate that organisms do not inherit characters or even phenotypes. What are inherited are genes and chromosomes. How a particular gene is interpreted into a phenotype often depends upon the condition of several other genes and the environment in which all those genes must function.

Summary

1 / *In sexually reproducing species two individuals are unlikely to be identical.*

2 / *Offspring tend to resemble their close relatives more closely than they resemble other members of the species.*

3 / *The basic patterns of heredity were discovered by Gregor Mendel in 1865 after working with several strains of true-breeding garden peas.*

4 / *Diploid organisms have two genes for each observable characteristic.*

5 / *Genes may exist in several forms (called alleles), which correspond to different appearances of the characteristic (the phenotype).*

6 / *Genes separate during gametogenesis so that each haploid gamete contains one gene for each characteristic.*

7 / *Fertilization that produces a new diploid generation restores the condition of two genes for each characteristic.*

8 / *Because recessive phenotypes do not occur in the F_1 but do occur in the F_2, the genes must be particulate structures.*

9 / *Monohybrid crosses with dominance always produce a 3:1 phenotypic ratio in the F_2.*

10 / *With intermediate inheritance the heterozygote always has a phenotype that is distinct from (and usually in between that of) the two homozygous phenotypes.*

11 / *The results of a test cross can be used to determine the genotype of an individual with a dominant phenotype.*

12 / *Mendel's studies of dihybrid crosses demonstrate that the alleles of genes for different characteristics separate into gametes independently of each other. Thus the dominant allele of one gene has no influence on the movement of the dominant (or recessive) allele for some other gene.*

13 / *Dihybrid crosses with dominance and no linkage always produce a 9:3:3:1 phenotypic ratio in the F_2.*

14 / *The phenomenon of independent assortment of alleles from different genes allows us to predict the phenotypic ratio in the F_2 of crosses involving multiple genes.*

15 / *Chromosomes may be viewed as a sequence of genes. Genes located on the same chromosome are said to be linked.*

16 / *Linked genes may be determined most easily be deviations from the expected equal frequency of recombinant and parental phenotypes in the offspring of a dihybrid test cross.*

17 / *The frequency of recombinant phenotypes from a dihybrid test cross is used as a measure of the distance between those linked genes on the chromosome.*

18 / *The occurrence of nonhomologous, paired sex chromosomes appears to be responsible for sex determination in some kinds of animals.*

19 / *Genes associated with sex chromosomes will exhibit different phenotypic ratios in reciprocal crosses.*

20 / *Two or more genes may affect the same characteristic.*

21 / *Phenotypic expressions of a gene can be modified by environmental conditions.*

Questions to Think About

The easiest way to demonstrate that you understand the concepts discussed in this chapter is to solve genetic problems. Biology professors have used this approach for testing students' knowledge for years. Repeated application of the abstract principles to realistic situations is the most highly recommended way of learning this material. The problems included here are a way of giving you some experience in this area as well as a chance to test yourself.

1 / There is a simple gene in chickens that controls the growth of feathers on the feet. The feathered allele is recessive to the naked (unfeathered) allele. If a rooster homozygous for feathered feet were mated to a hen homozygous for unfeathered feet, what genotypes and phenotypes would you expect to find in the F_1? In the F_2?

2 / Suppose you crossed a black Laborador Retriever with a blond Laborador Retriever and obtained nine puppies; four black and five blond. What could you conclude about the genetics of coat color in these dogs? What are the most probable genotypes for the different dogs? How could you clear up any uncertainties that you may have about this situation?

3 / If you grew 36 pea plants from peas obtained on a single plant and 27 produced red flowers while the remainder produced white blossoms, what is the most likely genotype of the parent plant? What would have been the phenotype of this plant?

4 / How many genotypically distinct kinds of haploid gametes can be produced by an individual with the diploid genotype *DD*? *Dd*? *dd*?

5 / What would you get if you tried crossing a fox with a chicken?

6 / List the genotypes and the proportions you would expect to obtain from the following crosses: (a) *Bb* × *bb*; (b) *Bb* × *Bb*; (c) *BB* × *bb*; (d) *BB* × *Bb*.

7 / If you cross-pollinate pink snapdragons, the offspring produce blooms that are either red, pink, or white in a ratio of 1:2:1. How do you account for this?

8 / In fruit flies there are two unlinked genes controlling body color and wing shape. The gray allele is dominant to the black allele for body color, and the straight allele is dominant to the curly allele for wing shape. If you crossed a fly homozygous for gray body and straight wings with one homozygous for black body and curly wings, what genotypes and phenotypes would you find in the F_1? What phenotypic ratios would you expect to find in the F_2?

9 / How many different kinds of gametes can be produced by an individual with the genotype *AABb*? *AABB*? *AaBb*? *AaBB*?

10 / Give the phenotypic ratios you would expect from the following crosses: (a) *LlQq* × *llqq*; (b) *LLqq* × *llQQ*; (c) *LlQq* × *LlQq*; (d) *Llqq* × *llQq*.

11 / In the F_2 of a dihybrid cross, what proportion of offspring would you expect to show both characters in the recessive condition?

12 / How many different phenotypes can be obtained from the offspring of the following cross: *AaBbCcDdEe* × *AaBbCcDdEe*?

13 / How many different kinds of gametes can be produced by an individual with the genotype *KKllMmnnOOPPqqRr*?

14 / What is the most likely phenotypic expression of color in the fruit of a ripe orange?

15 / In the cross *DdEe* × *ddee*, the following results were obtained: 37*DdEe*, 13*Ddee*, 11*ddEe*, and 35*ddee*. Are these genes linked? Why or why not? If they are linked, how many map units separate them?

16 / In the cross *NnRr* × *nnrr*, the following results were obtained: 42*NnRr*, 45*Nnrr*, 39*nnRr*, and 41*nnrr*. Are these genes linked? Why or why not? If they are linked, how many map units separate them?

17 / Color blindness is inherited as a sex-linked recessive condition. Two people with normal color vision marry and produce four children; three daughters and one son. The son is color-blind, but his three sisters have normal color vision. How can this happen?

18 / If a couple already has four daughters, what is the probability that their next child will be a son?

19 / In chickens there is a gene for feather color whose dominant allele (*C*) produces black color while the recessive allele (*c*) produces white color. There is another gene in chickens that, when present in the homozygous recessive condition (*ii*), prevents the production of dark feathers regardless of the condition of other genes. In the cross *CcIi* × *CcIi*, what are the phenotypes of these birds, and what are the predicted proportions and phenotypes of their offspring? Devise a means by which you could cross two white chickens and produce all-black offspring.

20 / Repeated crosses between hairless breeds of dogs and hairy breeds result in offspring showing the hairy and hairless phenotypes in a 1:1 ratio. Hairy dogs crossed with other hairy dogs always produce only hairy offspring. However, hairless dogs crossed with other hairless dogs produce hairy and hairless offspring in a 1:2 ratio. How do you explain the genetics of these facts?

Answers

1 / All the F_1 would have the genotype *Ff* (heterozygous). Their phenotype would be unfeathered feet. The F_2 would have three genotypes, *FF*, *Ff*, and *ff* in the ratio of 1:2:1. Because the unfeathered allele is dominant, this would give a phenotypic ratio of three unfeathered to one feathered.

2 / Based on the limited facts available, all you can conclude for certain is that there are two alleles for the coat color gene and that one allele is probably dominant to the other. The 1:1 phenotypic results of the offspring suggest that this is a test cross involving a dog homozygous for the recessive allele and another that is heterozygous.

	c	*c*
C	*Cc*	*Cc*
c	*cc*	*cc*

The easiest way to clear up the mystery would be to cross one black puppy with another and to cross one blond puppy with another. The phenotype corresponding with a homozygous genotype should produce only one phenotype in its offspring, that looking just like its parents. The other cross (using heterozygotes) should produce two kinds of offspring in the ratio of 3:1 (3 dominants to 1 recessive).

3 / The parent pea plant was heterozygous for the flower color gene and produced red flowers.

4 / One; two; one.

5 / Frustrated! Alternative acceptable answers include, but are not limited to, a

dead chicken, no offspring, a paunchy fox, or a lot of furious running around. Chickens cannot be crossed with foxes.

6 / (a) 1*Bb* :1*bb*; (b) 1*BB* :2*Bb* :1*bb*; (c) all *Bb*; (d) 1*BB* :1*Bb*.

7 / Flower color in snapdragons is an example of intermediate inheritance and pink flowers are heterozygous.

8 / All the F_1 would be heterozygous for both genes (*BbCc*), meaning they would have gray bodies and straight wings. In the F_2 you would expect $\frac{9}{16}$ to have gray bodies and long wings, $\frac{3}{16}$ to have black bodies and long wings, $\frac{3}{16}$ to have gray bodies and curly wings, and $\frac{1}{16}$ to have black bodies and curly wings.

9 / Two; one; four; two.

10 / (a) 1:1:1:1; (b) all the same (both characters in the dominant condition); (c) 9:3:3:1; (d) 1:1:1:1.

11 / $\frac{1}{16}$ or 6.25%.

12 / 32.

13 / 4.

14 / Orange.

15 / Yes, these genes are linked because there are not an equal number of parental types and recombinant types in the offspring. If the genes were not linked, you would expect equal numbers of all four phenotypes. These genes are separated by 25 map units.

16 / No, these genes are not linked because all four phenotypes occur equally in the offspring. This is an almost perfect 1:1:1:1 phenotypic ratio that you would expect only if the two genes were assorting themselves independently.

17 / The recessive allele for color blindness is carried on the X chromosome. The boy has only one X chromosome, which he obtained from his mother. This means the mother must carry the color-blind allele, but she is not color-blind. Women have two X chromosomes, meaning the mother's second X chromosome must have a dominant allele for normal color vision. The father has one X chromosome and is not color-blind, so it must carry the allele for normal color vision. All the daughters have normal color vision because they all carry an X chromosome obtained from their father. Because the allele for normal color vision is dominant to the allele for color blindness, even if they inherited the allele for color blindness from their mother, it would be masked by the allele from their father.

18 / $\frac{1}{2}$ or 50%.

19 / Both parental birds are black; 9 black:7 white; *ccII* (white) × *CCii* (white) → *CcIi* (black).

20 / The simplest explanation involves only one gene with intermediate inheritance and one of the homozygous conditions being lethal. The one-to-one ratio obtained from the hairy × hairless cross tells you that this is a test cross using a heterozygote. Since hairy × hairy always produces only hairy offspring, we must conclude that hairy dogs are homozygous for the hairy allele. That makes the hairless condition heterozygous. A cross using two heterozygous (hairless) dogs would be expected to produce a 1:2:1 ratio of homozygotes that are hairy, heterozygotes that are hairless, and some homozygous genotype that has a new phenotype. The third genotype causes death during the developmental stages before birth and is called a lethal genotype.

6

Molecular Basis of Heredity

Introduction

The patterns of heredity discussed in Chapter 5 are repeatable descriptions of observable reality. The consistency of these observations enables us to infer the existence of the gene as some kind of unit of heredity. Up to this point the gene has been an abstraction, a mythical entity whose existence is accepted as the simplest explanation of the observed patterns.

The patterns of heredity observed in classical genetics and the patterns of chromosomal movement during cellular replication are strikingly parallel. These parallel observations suggest that genes and chromosomes are closely allied. The phenomenon of linkage encourages us to believe that chromosomes are linear packages of genes somewhat analogous to a series of pearls or beads on a string.

Knowing that such entities exist, and being able to explain hereditary outcomes, constitutes an increase in our understanding of life. At the same time, this new knowledge raises even more questions about the nature of life and the reproductive process. What is a gene? How and why is it capable of replicating itself? And how does it control or determine the phenotype of the organism? It should be clear that a full understanding of genetics and reproduction requires a knowledge of the molecular structure and functioning of the gene. The gene must be viewed less abstractly and more materialistically.

The Identity of the Gene

Even before the world was aware of the existence of genes, people were attempting to identify the material of which living things are made.

In 1869 a Swiss chemist named Friedrich Miescher determined that the nuclear material was composed of protein and some other nonprotein substance. Because this other substance would not be readily identified as any of the standard kinds of organic compounds known at that time, it was given a new name. It could be found in nuclei and had pronounced acidic qualities, so it was named nucleic acid. (It is an amazing characteristic of humans that the ability to apply a name to something that is otherwise totally unknown somehow gives them the feeling that they now know something about it.) Miescher also determined that nucleic acid was composed of carbon, hydrogen, oxygen, nitrogen, and phosphorus, the last being an element not found in proteins or carbohydrates.

By 1924 a German chemist named Robert Feulgen had developed a specific staining technique for nucleic acid; it would produce a bright red color in the presence of nucleic acid but would otherwise remain colorless. When applied to eucaryotic cells, Feulgen's stain not only caused nuclei to turn bright red but revealed that the chromosomes of replicating cells were also composed of nucleic acid. This color reaction caused the introduction of another new word, *chromatin*, which is essentially synonymous with nucleic acid.

The combined work of Miescher and Feulgen had narrowed the search for the gene's identity to one of two candidates, protein or nucleic acid. Both kinds of chemicals had advocates claiming evidence to support the superiority of their chemical as the gene. For nearly thirty years the situation remained ambiguous. The ambiguity was enhanced by an extremely limited knowledge of the molecular structure of these organic compounds. Both classes of compounds were known to be extremely large molecules, probably composed of millions of individual atoms. Furthermore, these classes of compounds exist in many different forms, and the amount of diversity shown, particularly among the proteins, is overwhelming. Finally, both kinds of molecules are, chemically speaking, rather delicate. They are not easy to work with in a test tube. They are difficult to isolate. They have a strong tendency to stop doing whatever it is they do when removed from their normal cellular environment.

ENZYME DEFICIENCY DISEASES

Some of the early geneticists observed a strong connection between genes and proteins. A British physician named Archibald Garrod noticed that several human conditions appeared to be inherited as simple Mendelian recessive traits. One of these conditions is known as **alcaptonuria.** The urine of people having this condition turns black on exposure to air. Such people also are characterized by having unusually hard and dark **cartilages.** They are, in all other respects, perfectly normal, healthy people. Garrod found that the cause of the black urine in people with alcaptonuria was the presence of **homogentisic acid** (also called **alcaptone**) in their urine, a substance not found in the urine of people without the disease. Normally, homogentisic acid is an intermediate product in the conversion of the amino acid tyrosine to carbon dioxide and water. The critical conversion from homogentisic acid to acetoacetic acid is mediated by a specific enzyme that alcaptonurics do not have. People with the dominant allele for the gene have the enzyme that converts homogentisic acid to acetoacetic acid and thus have normal-colored urine. People homozygous for the recessive allele have no enzyme and thus homogentisic acid is not converted to acetoacetic acid; it remains in the urine and turns black when exposed to air (see Figure 6.1).

The simplest interpretation of these and many other enzymatic

Figure 6.1

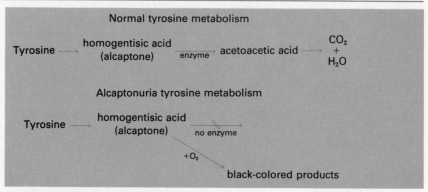

Tyrosine metabolism in humans.

differences with a genetic basis is that the enzymes are the genes. If you have the appropriate enzyme, you have one allele. If you have a non-functioning enzyme (or no enzyme), you have another allele. All enzymes are proteins. Enzymes are just a particular group of proteins that speed up the rate at which specific chemical reactions occur.

There was also a strong feeling among many biologists, particularly during the 1940s and early 1950s, that proteins were the only organic compounds complicated enough to confer the high level of specific diversity needed to make genes. Each species of organism has a very large number of different genes, each of which has several allelic forms. The known diversity of proteins can easily accommodate this diversity, but it was not clear that there were enough different kinds of nucleic acids.

THE CONSTANCY OF GENES, CHROMOSOMES, AND DNA

Subsequent workers provided many more examples of this already noted phenomenon, namely, the expression of phenotypes correlating with the structure of specific proteins. However, the genes-are-protein people did have a great deal of difficulty with amounts of protein. As we have already seen, the original studies of both classical genetics and cellular replication had independently established consistency in the amounts of genetic material present in the cells. All diploid cells of the same species must have two of each kind of chromosome and two of each kind of gene, but haploid cells of that species must have one of each kind of chromosome and one of each kind of gene. The immediate implication of this observation is that the material of which genes are made must be present in constant amounts in all diploid cells of a spe-

Table 6.1
DNA Content in Different Cells and Species

SPECIES	HAPLOID[a]	DIPLOID[a]
Cow	3.3	6.4–6.8
Chicken	1.3	2.4–2.6
Shad	0.91	1.99
Carp	1.6	3.0–3.3
Rainbow trout	2.45	4.9
Toad	3.7	7.3

[a] All values are given in picograms, which are equal to 10^{-12} grams.
Source: from Ruth Sager and Francis J. Ryan, *Cell Heredity,* New York, Wiley, 1961, Table 7.1, p. 178.

cies and halved in the gametes. Proteins do not fulfill this expectation. In fact, protein quantities often differ drastically in the same cell from one minute to another. In addition, the kinds of proteins found usually vary with the kind of cell examined. Finally, there is seldom any consistency in the amounts of protein found in diploid and haploid cells.

In contrast, once the techniques were worked out for quantifying the amounts of nucleic acid present in cells, all diploid cells of the same organism were found to contain the same amounts of DNA, which was twice the amount found in haploid cells of the same species (see Table 6.1). Here, then, was fairly strong evidence indicating that nucleic acid was the material of which genes are made. Few people were totally convinced without more evidence. This additional evidence was eventually found in a series of observations and experiments with bacteria and viruses.

BACTERIAL TRANSFORMATION AND VIRUS INFECTIONS

One of the practical benefits of early work with microscopes and the subsequent controversy over the cell theory and spontaneous generation (see Chapter 2) was the realization that many diseases were caused by microscopic cellular organisms. One such disease is a form of pneumonia caused by a bacterium called *pneumococcus*. By the late 1920s it was known that two different strains of the pneumococcus organism existed; each one had a different appearance under the microscope and produced a different effect when injected into living mammals. In one of these strains the microorganism was surrounded by a smooth, gelatinous covering; thus it was called the smooth, or S, strain. The other strain lacked the jellylike covering, giving it a much more textured appearance; thus it was called the rough, or R, strain. Injection of the smooth (S) strain pneumococcus into live mice invariably produced all the symptoms of pneumonia in these mice, resulting in their death, but the injection of the R strain pneumococcus did not. Furthermore, if one killed S strain pneumococcus by heating and then injected these dead bacteria into live mice, the mice did not get sick and die.

These results are very easy to interpret. Live pneumococcus with the genotype that produces the smooth, gelatinous covering are necessary to cause the disease pneumonia. However, the next observation is less easy to understand. An Englishman named Fred Griffith reported in 1928 that when he injected both heat-killed, S strain and live, R strain pneumococcus into his mice, they died of pneumonia. Neither of these factors could cause pneumonia alone. A postmortem examination of the mice revealed living S strain pneumococcus. After carefully repeating this experiment (illustrated in Figure 6.2) several times, Griffith was convinced that somehow the R strain pneumococcus were being transformed into S strain bacteria by the material from the heat-killed S strain.

It was soon discovered that **transformation** of R strain pneumococcus to S strain pneumococcus could be accomplished without the use of mice; transformations could be done in test tubes containing the proper growth medium. Furthermore, it became clear that a water-soluble extract of heat-killed S strain could cause transformation of R strain pneumococcus just as effectively as whole, killed bacteria. With that much of a clue, all that was necessary was to characterize all the different sorts of water-soluble components that could be isolated from heat-killed, S strain pneumococcus and determine which ones would transform R strain to the virulent form. In 1944 a trio of American workers led by O. T. Avery established that only the nucleic acid DNA caused the R strain to change its genetic characteristics. This is a strong line of evidence favoring DNA as genes.

Another supplement of information came from a group of viruses that infect bacteria. These are called **bacteriophages** or **phages.** Phages are composed of nothing but proteins and nucleic acids, and for the most part the nucleic acid is located in the core of the phage surrounded by a shell of protein. Phages attach themselves to the outside of a bacterium and apparently cause the bacterial cell to do nothing but manufacture new phages, at the expense of the bacterium's normal functioning. After a suitable lapse of time the bacterium bursts open, releasing hundreds of new phages suitable for infecting new bacteria. It was inferred that the phage injected something into the bacterium, which made it behave in this self-destructive manner. The obvious question is, What gets injected into the bacterium by the phage—nucleic acid or protein?

Taking advantage of the facts that nucleic acids contain phosphorus while proteins do not and that proteins contain sulfur while nucleic

Figure 6.2

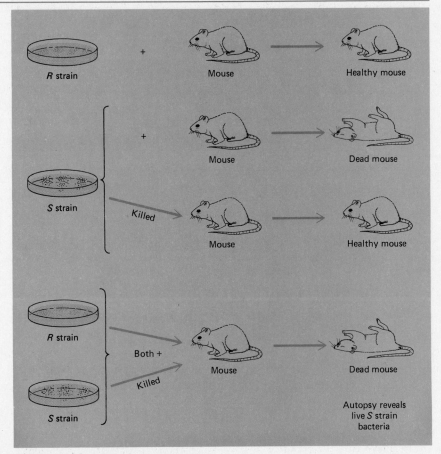

Genetic transformation in pneumococcus.

acids do not, a pair of American investigators, Alfred Hershey and Martha Chase, used radioactive phosphorus and sulfur to label the nucleic acids and proteins, respectively, of bacteriophages. These labeled phages were then allowed to infect bacteria, and the injected material was identified as containing radioactive phosphorus but no radioactive sulfur. This experiment, performed in 1952, showed finally that the genetic material was definitely nucleic acid and not protein (see Figure 6.3).

If the evidence overwhelmingly supports the nucleic acids as the substances of which genes are made, then what is the relationship of the various kinds of proteins to genes? We have already shown an extremely strong correlation between the presence of specific kinds of

Figure 6.3

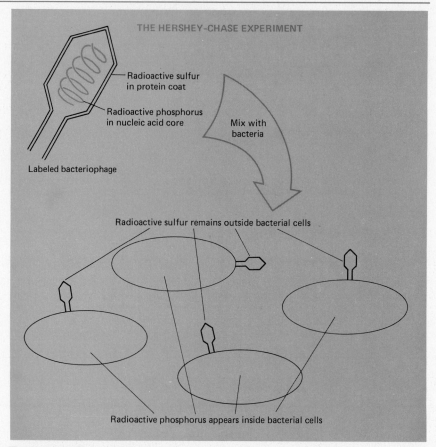

THE HERSHEY-CHASE EXPERIMENT

Radioactive sulfur
in protein coat

Radioactive phosphorus
in nucleic acid core

Mix with
bacteria

Labeled bacteriophage

Radioactive sulfur remains outside bacterial cells

Radioactive phosphorus appears inside bacterial cells

The Hershey-Chase experiment.

proteins and many hereditary phenotypes. The simplest working hypothesis is that proteins are the molecular equivalent of the phenotype of an organism, while nucleic acids are the molecular equivalent of the genotype. It would seem that the expression of a heritable characteristic in the genotype requires a conversion from a message encoded in the gene structure of DNA to a phenotypic translation incorporated into the structure of a protein molecule. In molecular terms what we need to ask is, How does a gene work to produce a phenotype? How does DNA control the production of protein? Obviously, one thing that would be of assistance in answering this question is a clearer knowledge of the basic structures of both DNA and protein.

Structure of Nucleic Acids

Nucleic acids are extremely large molecules called **macromolecules.** As is true for most macromolecules, nucleic acids are **polymers**—that is, they are composed of a larger number of simpler repeating units attached to one another in some sequential pattern. The monomers—the repeating units—of nucleic acids are called **nucleotides.** Each nucleotide consists of the union of three different kinds of molecules: an inorganic phosphate, a **pentose** or five-carbon sugar, and a complicated carbon-nitrogen molecule called a **nitrogenous base** (see Figure 6.4). The sugar found in DNA is always a pentose called deoxyribose. The sugar and phosphate of one nucleotide can form a bond with the sugar and phosphate of other nucleotides, thus forming a long sugar-phosphate chain. The nitrogen bases are attached only to the pentose molecules, so that a nucleic acid can be visualized as a long string of alternating sugars and phosphates with arms of nitrogen bases sticking out at regular intervals (see Figure 6.5).

In DNA there are four different nitrogenous bases that can be used in a nucleotide. These four different bases are **adenine, cytosine, guanine,** and **thymine,** and they are usually just referred to by their initial letters (see Figure 6.6). In the sense that there are four different bases that can be used to make a nucleotide, there are then four different kinds of nucleotides that can be used to make a nucleic acid. If a nu-

Figure 6.4

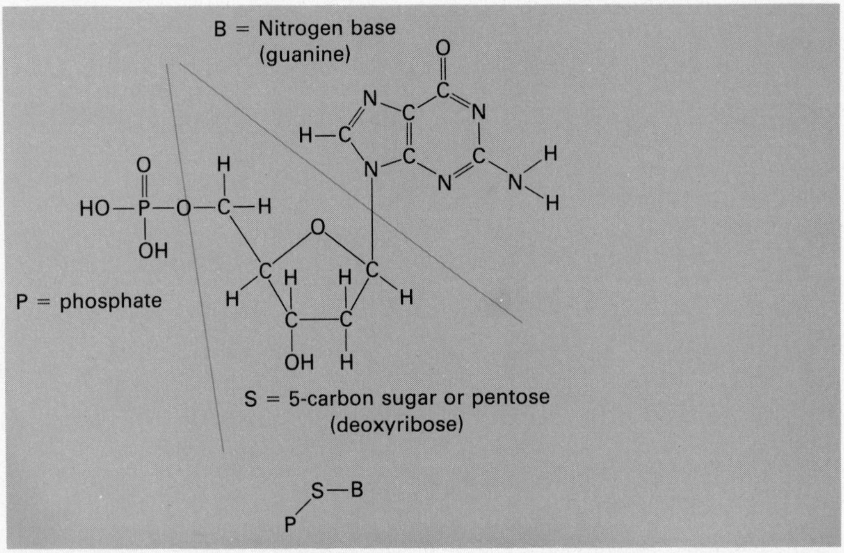

A nucleotide, the basic unit of DNA.

Figure 6.5

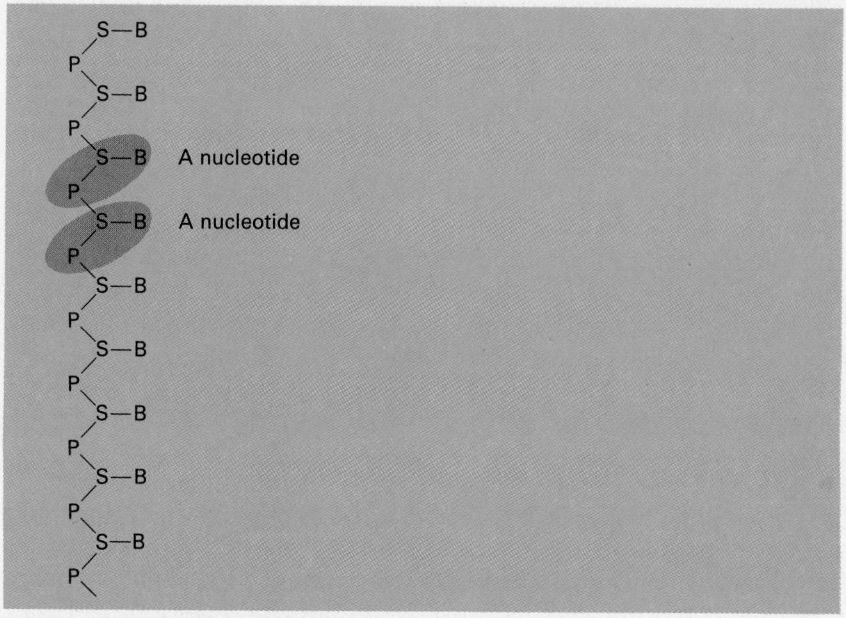

Elementary structure of a DNA molecule.

Figure 6.6

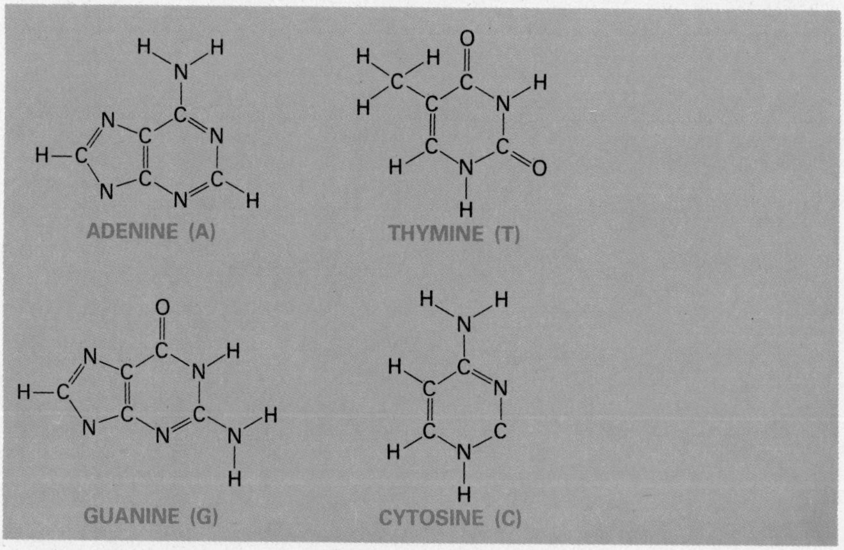

The four different bases found in DNA.

cleic acid is composed of several hundred nucleotides in sequence, it is possible to construct millions of different nucleic acids depending upon the sequence of the four specific nucleotides (see Figure 6.7).

To make matters more complex, it becomes evident that a chain of DNA is twice as thick as predicted on the basis of the foregoing description. Furthermore, chemical analysis had revealed that the proportion of the two bases adenine and guanine was always equal to the proportion of thymine plus cytosine. More precisely, the number of molecules of adenine in DNA was always equal to the number of thymine molecules, and the number of guanine molecules was always equal to the number of cytosines. These observations required James Watson,

Figure 6.7

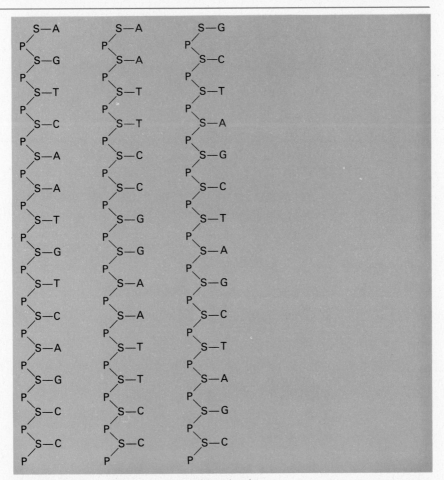

Examples of three different nucleic acid molecules.

Francis Crick, and Maurice Wilkins to conclude that DNA must consist of a double chain of nucleic acids whose bases are weakly bonded opposite one another in a specific, complementary pairing, A with T and C with G.

It is now possible to visualize a DNA molecule as a ladder whose two upright supports consist of alternating sugar and phosphate units and whose rungs consist of paired bases from opposite nucleotides. Each of the four bases has a different size, but the dimension of A and T placed together is equal to the distance spanned by C and G placed together. Thus only if adenine is opposite thymine, or cytosine is oppo-

Figure 6.8

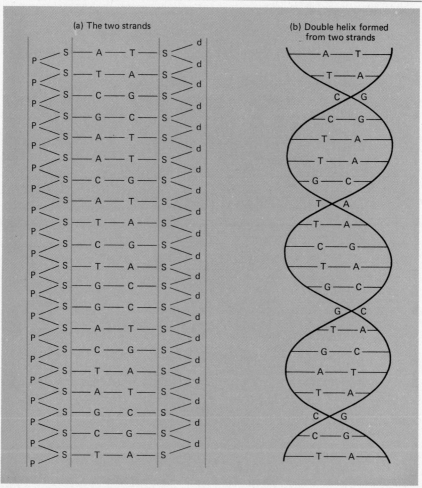

DNA is a double-stranded molecule.

site guanine, can the uprights remain equidistant and the rungs complete the gap between uprights. As a final twist, these two chains are actually coiled around one another in a **helical** fashion, as if our illustrative ladder were made of rubber, and while one end remained fixed, the other was rotated (see Figure 6.8).

It is particularly important to notice how the structure of this molecule can be used to explain the ability of genetic material to replicate itself. The very nature of the DNA molecule is such that it can be used as a model for its own duplication. The process is illustrated in Figure 6.9.

Figure 6.9

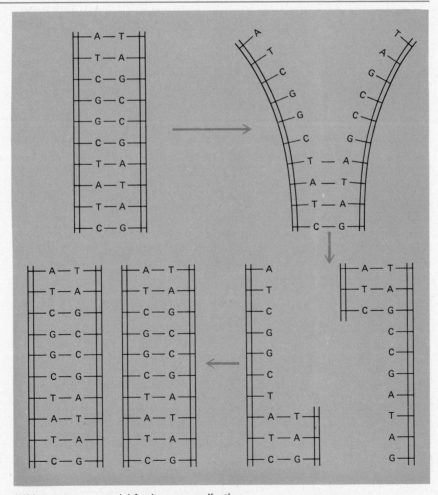

DNA serves as a model for its own replication.

If a double-stranded chain of DNA were to separate into individual strands, the complementary properties of the nitrogenous bases would place a restriction on the kinds of nucleotides that could join the isolated strands. A free-floating pool of nucleotides can react with each of the single strands, but only adenine-containing nucleotides can join to an exposed thymine, and only cytosine-containing nucleotides can join to an exposed guanine. What results from this process are two double-stranded molecules of DNA, each of which has exactly the same sequence of nucleotides (hence they are identical copies) and each of which has the same sequence of nucleotides found in the original DNA molecule. Such a process will require energy and several enzymes to facilitate its completion, but all the information needed to construct duplicate nucleic acids is present in the DNA molecule itself.

Structure of Proteins

Proteins are also macromolecules, and they are also polymers. The repeating subunit of a protein (or a **polypeptide**) is called an **amino acid.** All amino acids have the same basic construction, which is shown in Figure 6.10. However, they vary drastically with respect to what occurs in the position marked R. There are, quite literally, hundreds of different combinations of elements that can be placed in the R position of this kind of amino acid, and each of these thus constitutes a different amino acid. Fortunately, only about twenty of these R groups are commonly found in natural proteins. A sampling of some of these amino acids is illustrated in Figure 6.11.

Amino acids can be joined together if the **organic acid group** of one amino acid is allowed to react with the **amino group** of another, thus forming what is known as a **peptide bond** (see Figure 6.12). Two amino acids joined by a peptide bond are called a **dipeptide**, three are called a **tripeptide,** and many are called a polypeptide. A polypeptide may be a protein. On the other hand, two or more polypeptides may

Figure 6.10

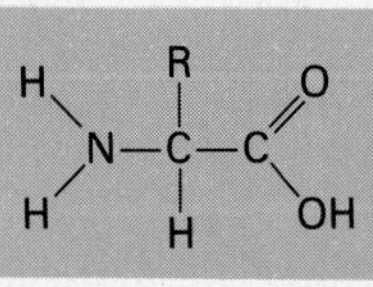

Diagrammatic representation of an amino acid molecule.

Figure 6.11

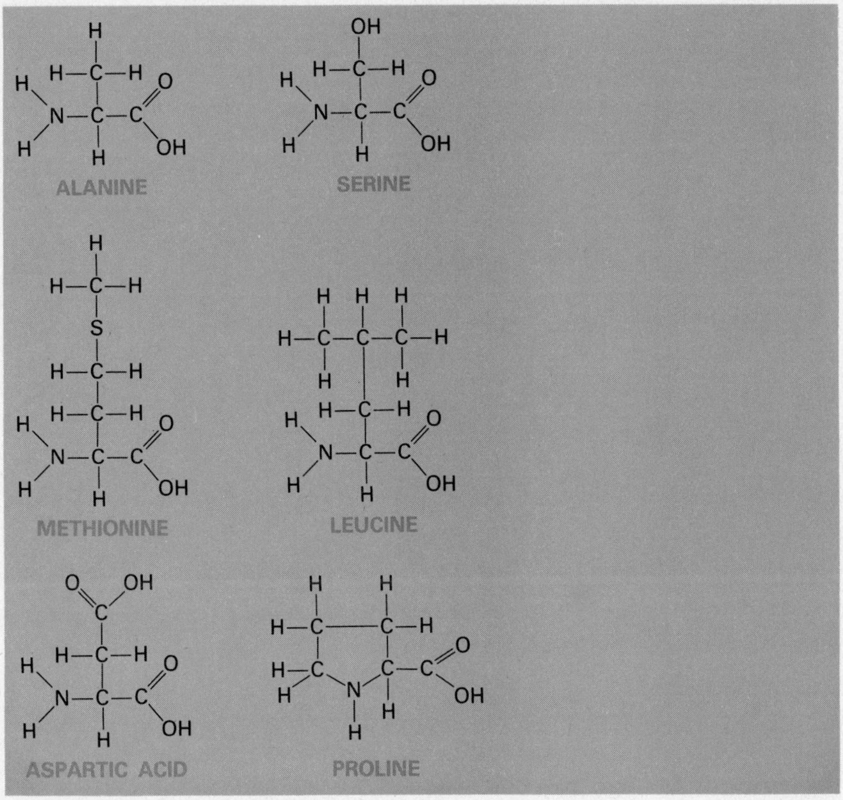

Some examples of amino acids.

have to join together to form a protein. A molecule cannot be called a protein until it is capable of performing some function. Thus if a single polypeptide is functional, it is called a protein, and if several polypeptides must combine to be functional, then they are the protein. (See Figure 6.13 for examples of proteins.)

It is easy to get the idea that proteins are just long strings of amino acids held together by peptide bonds. Unfortunately, each of the amino acids has slightly different physical properties, which usually result in the polypeptide being bent and twisted into a relatively compact structure. The three-dimensional configuration of each protein is unique and apparently crucial to its function but almost completely a result of the sequence of its constituent amino acids.

Figure 6.12

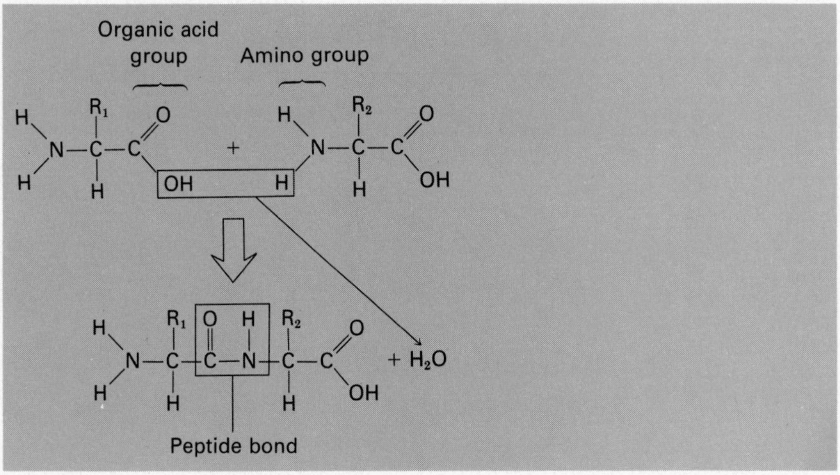

Formation of a peptide bond.

Relationship Between DNA and Protein

We have determined that DNA is a polymer whose repeating subunits are four different kinds of nucleotides and that proteins are polymers whose repeating subunits are twenty different kinds of amino acids. We also know that DNA is found almost exclusively in the nucleus but that proteins, while found in all parts of the cell, are manufactured only on the ribosomes. Because we know that the genotype (composed of DNA molecules) has to have an effect on the phenotype (composed of protein molecules), there are two obvious questions we should ask now. If the informational importance of both kinds of molecules is in the sequence of their subunits, is it possible that the sequence of nucleotides in DNA in some way affects the sequence of amino acids in proteins? Is there any physical connection between the nucleus where the DNA is retained and the ribosomes where the proteins are assembled?

The second question is the easiest one to answer. All it requires is high-level microscopic investigation of cell structure to see that the endoplasmic reticulum is a series of small tubes from the nucleus to the cell membrane, with many of the ribosomes attached along the way. The existence of this pipeline between the site of genetic information storage and the site where genetic information is translated into phenotype very strongly implies the existence of some kind of messenger

Figure 6.13

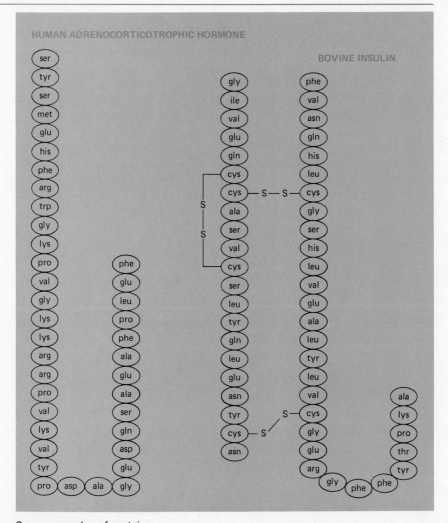

Some examples of proteins.

substance that travels from the nucleus to the ribosomes. If the sequences of both the nucleotides and the amino acids in their respective polymers are critical, then it only makes sense to assume that the messenger substance must also have some kind of sequential structure that will correctly and unambiguously transmit this crucial information.

STRUCTURE OF RIBONUCLEIC ACIDS

The most likely molecular candidates for this exacting messenger role are a group of nucleic acids known as **RNA** (for *ribonucleic acid*). RNA is also a polymer whose repeating subunit is a nucleotide. However, the pentose sugar in RNA is ribose (not deoxyribose as in DNA) and the base **uracil** occurs instead of thymine. Ribonucleic acids occur as single-stranded, helical coils instead of as double-stranded coils, and they can be found in the cytoplasm as well as in the nucleus. In fact, they are frequently found associated with ribosomes. The kind of RNA that travels from the nucleus to the ribosome is called (reasonably enough) **messenger RNA,** or mRNA.

In much the same way that DNA strands can serve as the basis for their own replication, they can also be used as the model for the synthesis of RNA molecules. The only significant difference is that when a nucleotide containing adenine is encountered in a DNA chain, it has to be complemented by a nucleotide containing uracil in the RNA chain. There is obviously some difficulty in determining how the cell knows which of the two chains of nucleotides in the DNA is to serve as the basis for synthesizing the mRNA, but it is certainly clear that, conceptually, DNA sequences can be used to determine RNA sequences.

SOLVING THE GENETIC CODE

Knowing that the nucleus uses an intermediary (mRNA) to direct the assembly of proteins provides a means by which the translation process, from nucleotide sequence to amino acid sequence, can be investigated. As we learned in Chapter 2, removal of the nucleus from a cell inevitably results in the cell running out of control. But it is possible to remove cytoplasmic constituents, such as ribosomes, and maintain them in a cell-free environment for an indefinite period of time in order to study their metabolic activities. This obviously requires some relatively sophisticated chemistry. If you've got one of these cell-free ribosomal systems set up with all the proper ingredients, it should be possible to give it one specific messenger RNA and have the system produce measurable quantities of one specific polypeptide. This experiment was done in the early 1960s, but the technology for working with both nucleic acids and proteins was so difficult that it was not possible to determine the relationship between the nucleotide sequence of the mRNA and the amino acid sequence of the resulting protein. In effect, something was being pushed into the system and something else was being pulled out, but the relationship between the two was obscure.

Once it was shown that a cell-free system would work to synthesize polypeptides, the next step was to supply such a system with a messenger RNA of known sequential composition, preferably one with a very simple composition. The ideal was achieved when several biochemists managed to construct a totally artificial mRNA molecule composed of nucleotides all containing the base uracil. When this poly-U mRNA was given to a cell-free system of ribosomes, it dutifully read the message and produced a polypeptide consisting of repeating units of the amino acid phenylalanine. Poly-U messenger RNA must specify or code for the production of polyphenylalanine. Similarly, when the cell-free ribosomal system was given an mRNA consisting of nucleotides that all contained the base cytosine (poly-C), a polypeptide was produced that consisted of the repeating amino acid proline. In addition, poly-A was found to produce polylysine, and poly-G encoded the production of polyglycine.

As a momentary aside, it should be very apparent that there cannot be a one-for-one correspondence between the nucleotides in the nucleic acids and the amino acids in the proteins. If such were the case, then it would only be possible to have 4 different amino acids found in polypeptides, the 4 (phenylalanine, proline, lysine, and glycine) that were found when using the 4 messenger RNAs composed of only one kind of nucleotide. But we have already noted that there are at least twenty different kinds of amino acids commonly found in polypeptides. This means that a single-nucleotide code unit is inadequate to account for the observed diversity found in polypeptide structure. In the same way, a pair of adjacent nucleotides would also be inadequate as a code unit because the maximum number of different code units that could be achieved by taking 4 different kinds of nucleotides, 2 at a time, would be 16. Thus the minimum requirement for a coding system based on sequential nucleotides of 4 different types is a coding unit that consists of 3 nucleotides. Such a coding system will allow for 64 different code units (or **codons**), which is more than enough for our requirements of approximately twenty (see Table 6.2).

By constructing artificial mRNAs with different proportions of the four different nucleotides, researchers have determined the coding system used in living material, as shown in Table 6.3. Notice that in many places the code is **redundant**—that is, many different coding units may specify the same amino acid. However, it is equally important to notice that the code is not **ambiguous**; there are no cases where a code unit can specify more than one amino acid. The redundancy feature of the

Table 6.2
Coding Systems for Proteins

One DNA nucleotide can code for 4 different amino acids

A	T	C	G

Two DNA nucleotides can code for 16 different amino acids

AA	TA	CA	GA
AT	TT	CT	GT
AC	TC	CC	GC
AG	TG	CG	GG

Three DNA nucleotides can code for 64 different amino acids

AAA	ACA	TAA	TCA	CAA	CCA	GAA	GCA
AAT	ACT	TAT	TCT	CAT	CCT	GAT	GCT
AAC	ACC	TAC	TCC	CAC	CCC	GAC	GCC
AAG	ACG	TAG	TCG	CAG	CCG	GAG	GCG
ATA	AGA	TTA	TGA	CTA	CGA	GTA	GGA
ATT	AGT	TTT	TGT	CTT	CGT	GTT	GGT
ATC	AGC	TTC	TGC	CTC	CGC	GTC	GGC
ATG	AGG	TTG	TGG	CTG	CGG	GTG	GGG

code accounts for the utilization of almost all the excess codons, those extra codons inherent in the triplet system beyond the twenty needed for each specific amino acid found in polypeptides. Three of the codons appear not to encode for any amino acid but specify instead that the polypeptide sequence be terminated at this point. This coding presents a situation analogous to the symbols used in language to denote punctuation. In effect, these three codons indicate the end of the polypeptide and signal the synthesizing machinery to turn off.

How does the sequence of nucleotide codons in a chain of messenger RNA cause the appropriate amino acids to join in sequence, forming a polypeptide? Or, more specifically, how do the twenty different kinds of amino acids recognize the appropriate codons calling for their inclusion at a specific locality? The answer to these questions lies in yet another form of nucleic acid, found as part of the protein-synthesizing machinery that was talked about earlier—the **transfer ribonucleic acids,** or tRNAs. There are at least sixty-one different tRNAs, one for each codon that specifies an amino acid. Each tRNA can be activated by a **catalyst** to selectively bind one molecule of the appropriate amino acid. On each tRNA is a sequence of nucleotides called an **anticodon.** This is a triplet of nucleotides that is complementary to the triplet codon carried by messenger RNA. As each codon on the mRNA is exposed at the ribosome, it attracts the anticodon of a tRNA carrying the appropriate amino acid. Successive, sequential additions of amino

Table 6.3

The Genetic Code in mRNA Codons

FIRST NUCLEOTIDE	SECOND NUCLEOTIDE[a]				THIRD NUCLEOTIDE
	U	C	A	G	
U	phe	ser	tyr	cys	U
	phe	ser	tyr	cys	C
	leu	ser	Stop	Stop	A
	leu	ser	Stop	trp	G
C	leu	pro	his	arg	U
	leu	pro	his	arg	C
	leu	pro	gln	arg	A
	leu	pro	gln	arg	G
A	ile	thr	asn	ser	U
	ile	thr	asn	ser	C
	ile	thr	lys	arg	A
	met	thr	lys	arg	G
G	val	ala	asp	gly	U
	val	ala	asp	gly	C
	val	ala	glu	gly	A
	val	ala	glu	gly	G

[a] asp = aspartic acid   met = methionine   arg = argenine   leu = leucine
    glu = glutamic acid   trp = tryptophan   cys = cysteine   pro = proline
    phe = phenylalanine   gln = glutamine   tyr = tyrosine   lys = lysine
    asn = asparagine   his = histidine   ala = alanine   ser = serine
    ile = isoleucine   thr = threonine   gly = glycine   val = valine
   Stop = termination of the polypeptide

acids to the structure are joined by peptide bonds, facilitated by the presence of enzymes. What results is a polypeptide chain that grows until a **terminator codon** is reached. At this point the completed polypeptide is released into the cytoplasm (see Figure 6.14).

Molecular Genetics in Perspective

The reaction of most students to this impressive array of minute chemical detail is one of utter bewilderment. What, if anything, has all this business about DNA, RNA, and proteins got to do with real genetics such as flower color in garden peas or eye color in humans? What is the relevance of all of these intricate activities? Perhaps the confusion can be removed by giving further consideration to the classic concept of the allele.

The allele, you will recall, was proposed as an alternative state in which a gene may exist, the gene being that bit of genetic material re-

Figure 6.14

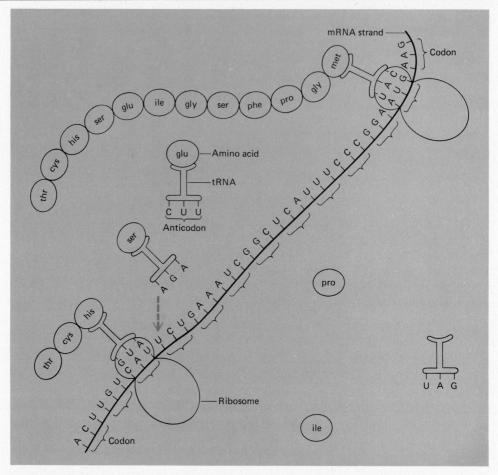

Protein synthesis.

sponsible for the production of a character. As an example, the gene for flower color in garden peas can exist in either of two allelic states, red or white. The phenotypic expression of these alleles can be described at several different levels. Flowers may appear either as red in color or as white. Alternatively, the same phenotypes can be described as possessing or lacking a specific red-colored pigment. At a third level of analysis the two phenotypes can be described as having a functional enzyme that results in the production of visible amounts of the specific red-colored pigment. If you look at the phenotype in this last way, then it is easier to visualize the gene as that piece of DNA responsible for encoding the amino acid sequence of the polypeptide that functions as the enzyme (or a part of the enzyme)that produces the flower color pigment.

What happens if somewhere in the life history of garden peas (in all those repeated mitoses and meioses) a mistake is made in the copying of a DNA strand that is ultimately responsible for flower color? The substitution of one nucleotide for another in a strand of DNA could easily result in the substitution of a different amino acid in the resultant polypeptide. A change in the structure of a polypeptide to even this minor a degree is often enough to change its functional characteristics to the extent that it may no longer produce the appropriate red pigment.

If such a nucleotide substitution has become incorporated into DNA structure, it is reasonable to assume that subsequent replications of the molecule will maintain this new nucleotide sequence. Thus all cells having this allele (this new form of the gene) will lack the ability to produce the original, functional enzyme and will pass this trait to their offspring. This incorporation of a heritable change into an organism constitutes a mutation.

There are several ways that mutations can occur; the present example of base substitution is perhaps the most obvious. It is also possible to delete a nucleotide or insert an extra nucleotide into a DNA sequence, thus totally altering the message within the sequence of triplet codons. Even grosser mutations can be caused by breaking, losing, or relocating large pieces of chromosomes. Many sorts of mutations can be induced by various sources of ionizing radiation, a multitude of chemical substances, heat, ultraviolet light, and possibly microwave radiation. It is worth noting that these diverse inducers of mutations cause random kinds of changes in DNA, so that it is not usually possible to produce a specific heritable change.

If such an event occurs in a diploid organism such as the garden pea (or human beings), the immediate effects would probably not be noticeable because there are two copies of each gene, one of which has presumably not been altered and continues to act as the basis for the production of functional enzymes. When one gene is responsible for producing a functional enzyme and the other is not, pigment will still be produced, so the flower petals will still appear red—maybe. We start now to approach an understanding of the molecular basis of dominance.

When one contemplates the events occurring at the cellular level of sexual reproduction, it is easy to see how an organism could be produced in which both copies of a gene have nucleotide sequences that code for the production of nonfunctional enzymes. The occurrence of this state of affairs in the flower color gene of garden peas results in the flower petals having a white color.

There is an interesting intellectual puzzle associated with the phenomenon I've just described. Obviously, red flowers are caused by the presence of red pigment. But white flowers are caused by a lack of red pigment in conjunction with a whole set of circumstances in the flower structure that are probably controlled by other genes. Consider for a moment the differences in the phenotypes between individuals heterozygous for flower color genes in garden peas and in snapdragons. As you will recall from Chapter 5, garden peas exhibit dominance but snapdragons do not. Thus heterozygous peas have red flowers, but heterozygous snapdragons have pink flowers. If in both cases flower color is caused by a pigment whose production is controlled by a protein molecule, then why should a single-dose production of enzyme in one case result in a red phenotype and in the other case result in a pink phenotype? One possible explanation depends upon the exact nature of the change in the mutated allele and thus the modification incorporated into the abnormal enzyme. It is also possible that the explanation for the difference resides not in the gene itself but in the background conditions within which the gene functions.

Assume for the moment that the substance(s) from which the red pigment is made has no color. In both peas and snapdragons the normal red allele for flower color controls the production of an enzyme that regulates the conversion of the colorless starting material into red pigment. It seems probable that in peas with the abnormal white allele the colorless starting substance is not converted to red pigment (nor is it necessarily converted into anything else). In the homozygous condition the flowers appear white because the structural configuration of the cells in the petals lacks pigment, and in the heterozygous condition enough enzyme is produced to convert some starting material to pigment so that the flowers look red. It is possible that in snapdragons with the altered flower color genes, the resulting enzyme is modified to the point where it converts the colorless substances to a white pigment. Thus in a heterozygous organism both white and red pigments are produced, giving a pink appearance. Alternatively, the white allele in snapdragons may only produce a nonfunctional enzyme, so that the heterozygous plant appears pink because of the interaction of a reduced amount of red pigment with a white-appearing structural background.

The foregoing discussion is admittedly highly speculative, but it is included for two important reasons. The first of these is to demonstrate that scientists cannot yet give completely factual explanations to every phenomenon that exists in our world. Despite the fact that we have been studying the genetics of garden peas for over a hundred years, we do not yet know everything there is to know about the subject. The second reason for the lengthy discourse is that it is a reasonable example of the way scientists think in trying to formulate an explanation for the phenomena that exist in nature. There are obvious, isolated bits of information that all have a bearing on the genetics of flower color. What relationships, if any, can be found among them? If they can be pieced together into a coherent generalization, is it possible to predict certain outcomes and/or test the validity of the generalization?

Perhaps a clearer illustration of the relationship between classical genetics and molecular genetics can be found in the inheritance of sickle-cell anemia in humans. Sickle-cell anemia is an abnormal condition, found typically in blacks, characterized by acute crisis symptoms of severe pain in the limb joints and in the abdomen. The victims are frequently afflicted with localized sores and ulcerations, general weakness, and, occasionally, coma. Routine medical examinations often reveal kidney and liver damage, an enlarged heart, characteristic changes in the shape of some limb bones, and traces of blood in the urine. People with this condition have a greatly reduced life expectancy. The most distinctive feature of this disease is the presence of abnormally shaped cells in the blood of patients. Normal red blood cells are biconcave disks that resemble doughnuts [see Figure 6.15(a)]. Victims of sickle-cell anemia have a large number of elongated, crescent-shaped red blood cells that resemble the blades of a sickle [see Figure 6.15(b)].

This collection of characteristics for sickle-cell anemia appears to be inherited as a simple Mendelian trait. In examining the parents of 21 victims of sickle-cell anemia, James Neel, an American geneticist, found that all 42 had a small number of sickle cells but did not demonstrate any of the other characteristic features of the anemia. He therefore concluded that they must be heterozygous for the condition. The major difference between people homozygous for the sickle-cell condition and those heterozygous for the sickle-cell trait is in the relative number of abnormal red blood cells. It appears as if all the other characteristics of sickle-cell anemia are due to the large number of abnormally shaped cells getting stuck in small blood vessels. This deprives the living cells beyond the blockage of many needed resources distributed by the blood, most particularly oxygen.

From what has been said earlier about the relationship between genes and phenotype, you should suspect that the significant differences between people with normal red blood cells and those with sickle cells must involve some difference in a kind of protein. A study

Figure 6.15

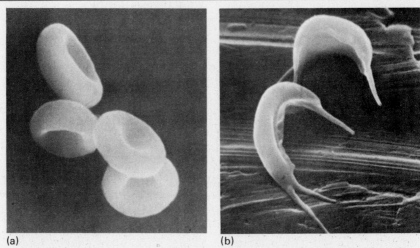

(a)   (b)

Normal (a) and sickled (b) red blood cells. From Robert M. Chute, *Introduction to Biology*, New York, Harper & Row, 1976, Figures 1-1c and 1-1d. [Photos courtesy of Dr. J. G. White, University of Minnesota Medical School.]

of normal red blood cells reveals that they contain large amounts of an oxygen-carrying molecule called **hemoglobin.** Hemoglobin is a complex structural entity consisting of four iron-containing **heme** groups held together by four polypeptide chains. In normal, adult, human hemoglobins there are two different kinds of polypeptides: two alpha chains and two beta chains (see Figure 6.16). Using rather complex, chemical methods, one can compare the hemoglobin molecules of normal adults and those with sickle cell. The only significant difference between the two is found in the beta chains. The beta chain is a polypeptide consisting of 146 amino acids. In normal hemoglobin there is a glutamic acid in position No. 6, but in sickle-cell hemoglobin there is a valine at this site. The single substitution of an amino acid in position 6 of the beta chains is sufficient to reduce the oxygen-carrying capacity of the whole hemoglobin molecule and alter its three-dimensional configuration in such a way that it can complex into tubular structures with other such modified hemoglobins. These tubular complexes of modified hemoglobins are what produce the distorted shape of the red blood cells and result in the clinical symptoms of the anemia.

An examination of the genetic code in Table 6.3 reveals that the mRNA codons for glutamic acid can be either GAA or GAG, while the codons for valine can be GUA or GUG. It is easy to see that the substitution of a uracil-containing nucleotide for an adenine-containing nu-

Figure 6.16

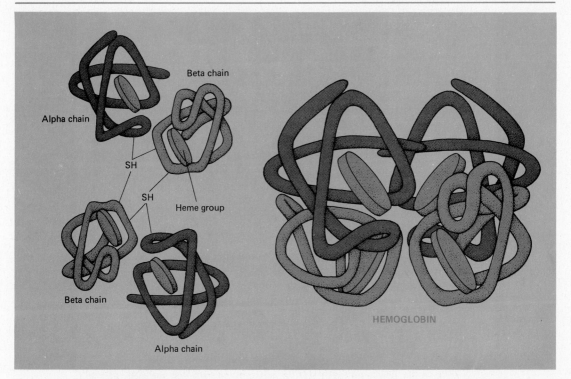

The hemoglobin molecule.

cleotide in the second position of either one of these mRNA triplets could change the amino acid specified in a polypeptide from a glutamic acid to a valine. More complex, and therefore less probable, substitutions could accomplish the same outcome.

It should be apparent that people suffering from sickle-cell anemia do not have a disease in the normal sense of the word. It is not possible to discover some medicine that will destroy the causitive agent of the abnormal symptoms, because this agent is the genetic basis of the person. Each and every cell of the person contains the instructions for building sickle-cell hemoglobin, encoded in the structure of the nuclear DNA molecules. It may be possible someday to discover a method by which the clinical effects of the altered beta chains are ameliorated so that the person can live a more normal life without the attendant pain and suffering. However, such a person cannot be thought of as cured, because he or she still carries the sickle-cell genes and can still pass them on to the next generation.

Summary

1 / *The center of cellular regulation and control is the nucleus.*

2 / *The nucleus is composed of protein and the nucleic acid DNA.*

3 / *DNA can be found as diffuse chromatin in the nucleus or in chromosomes.*

4 / *The expression of genetic characteristics is determined by the presence or absence of particular proteins.*

5 / *All diploid cells of the same organism have a constant amount of DNA, which is double the amount found in haploid cells of that species. Amounts of protein found in cells of a species are inconsistent with any genetic expectations.*

6 / *Alteration of genetic characteristics, as in bacterial transformation or bacteriophage infection, can be brought about only by DNA, not by protein.*

7 / *DNA is the genetic material. Proteins must be the phenotype at the molecular level of analysis.*

8 / *DNA is a polymer whose repeating subunit is a nucleotide. A nucleotide is composed of a pentose, a phosphate, and a nitrogenous base. Any one of four different nitrogenous bases (A, T, C, G) can occur in the nucleotide of a DNA molecule. Two chains of nucleotides are coiled about one another at a constant distance maintained by complementary paired nitrogenous bases (A with T and C with G). The sequence of the four different kinds of nucleotides contains information.*

9 / *The complementarity of the two nucleotide strands in DNA is the informational basis for the exact replication of the DNA molecule.*

10 / *Protein (or polypeptide) is a polymer whose repeating subunit is an amino acid. Any one of twenty different kinds of amino acids can be found in the structure of proteins. The sequence of the twenty different kinds of amino acids contains information and determines the specific characteristics of the protein.*

11 / *The genetic information encoded in the sequence of nucleotides of DNA is translated into the sequence of amino acids of protein by way of intermediary nucleic acid molecules called mRNA and tRNA.*

12 / *A sequence of three nucleotides in a nucleic acid (a codon) specifies a particular amino acid.*

13 / *The difference between normal hemoglobin and that found in people with the Mendelian recessive condition called sickle-cell anemia is in the substitution of one amino acid out of 146 in the beta chains of the protein.*

14 / *A mutation is any physical change in the structure of a DNA molecule.*

Questions to Think About

1 / What happens to the hair color gene of a man who goes bald?

2 / What does the flower color gene do before the plant produces flowers?

3 / Do the answers to the first two questions give you any insight on how cellular differentiation may occur?

4 / How many different genes are involved in producing a human phenotype? How many different polypeptides are needed? How big is each of the kinds of

molecules? How many mistakes can be made? What is a phenotypically normal person?

5 / Why are some alleles dominant to others? Is the allele for normal hemoglobin dominant to the allele for sickled hemoglobin?

6 / How many different ways can you think of to alter the phenotype of an individual without changing his or her genotype? Has anyone actually changed his or her phenotype? What are the purposes of wigs, hair dyes, cosmetics, colored contact lenses, and so forth?

7 / How would you go about eliminating a genetic disease such as sickle cell anemia?

7

Changes in Hereditary Makeup

Introduction

One of the truly impressive observations made by almost anyone who attempts to study living things is the incredible amount of diversity exhibited by living forms. Perhaps the greatest handicap to making simple generalizations in biology is the huge number of different kinds of organisms. More often than not a generalization must be qualified to account for some exceptional organism that doesn't quite fit the rule. The difficulties seen in describing life or the appearance of the cell are examples that illustrate this almost overwhelming diversity. The diversity of life in itself constitutes a major generalization of biology. Living things exhibit many differences among themselves.

The most obvious kinds of diversity, of which we all become aware at a relatively early age, are the differences that can be found among species. Such differences are often so clear that they are the basis for applying different names. Dogs, daffodils, earthworms, and trout are all very different kinds of organisms. There are probably several million additional examples. The occurrence of so many differing species has encouraged many attempts at explanation.

It is a very old observation that the structures of different species of organisms enable those species to live in very particular ways (see Figure 7.1). The wings of birds enable them to fly, the long neck of the giraffe enables it to feed in the upper reaches of trees, and the fins of fish enable them to swim rapidly. This close correlation between the structure of a species and the way it lives has led many people to believe that there must be some purpose at work that can account for the observations. Something must have made all those different kinds of organisms the way they are so that they can do all the different things that they do.

In addition to the diversity that we can see among the species, there is also a great deal of diversity within each species (see Figure 7.2). Individual organisms frequently differ from one another even though they are members of the same species. Many of these differences can be explained by sexual condition, age, accident, or disease, but many of them cannot. The existence of individual variability was ignored for a long time or considered to be of no particular importance.

On the face of it, there is a strong element of inconsistency in these pronouncements about diversity. The differences between species appear important and imply some higher purpose, while the differences within species are unimportant and, at best, only imply that there are some imperfections in the mechanism of reproduction. This view of variability makes even less sense when you consider the amount of furor that has been raised by people over the differences that occur among people.

Figure 7.1

(a)

(b)

(c)

(d)

The structural features of many kinds of organisms are obviously related to the way those organisms live and the functions they perform. (a) Gannet in flight; (b) gourami; (c) giraffe; (d) rhinoceros. [Photo (a) courtesy of Dr. John B. Hess. Photo (b) courtesy of John E. Wiley. Photos (c) and (d) by the author.]

Figure 7.2

Careful inspection will reveal that no two of these cattle look exactly the same and that each can be recognized as an individual. Individual variability is a common phenomenon among living organisms. (Note that one of the cows is actually a horse.) [Photo courtesy of John E. Wiley.]

Individual Variation: A Puzzle

Consider the kinds of individual variabilty with which we are familiar in the human species. With the possible exception of identical twins, no two humans are ever exactly alike (see Figure 7.3). Even in the case of identical twins, people who know them very well are usually able to differentiate them. Looking at another individual human being is not the same as looking into a mirror. Individual people differ in a multitude of ways, including height, weight, eye color, hair color, skin color, proportional size of limbs, nose shape and size, ear shape and size, distribution of hair on various body parts, and so forth and so on. People who have worked extensively and closely with other species of organisms are capable of distinguishing individuals easily.

As I implied in earlier chapters, much of the variability within a species has a genetic basis. Many genes exist that are responsible for the production of phenotypic characteristics. We have seen, in Chapter 6, that a chemical basis exists that enables genes to assume different allelic forms. Different alleles of a gene can produce different phenotypes, meaning that different individuals with different genotypes will be different.

SOURCES OF GENETIC VARIABILITY

As you could probably guess, any change in the molecular structure of the genetic material, the nucleic acids, would constitute a mutation and would be a source of individual genetic variability. Generally, in-

Figure 7.3

Further examples of the kinds of variability found among individuals. [Photos courtesy of Matthew E. Stockard.]

dividuals with different genotypes have different phenotypes. The biggest exception to that generalization occurs when dominance exerts an influence in heterozygous genotypes. If all genes are able to mutate independently, it is possible to generate an extremely large number of different phenotypes among a sufficiently large number of individuals.

If you add to the basic variability possible from mutations the reshuffling of genetic combinations that are produced by sexual reproduction, crossing-over, and chromosomal rearrangements, the array of different-looking phenotypes within a species becomes astronomical. This idea can be demonstrated mathematically.

For example, Mendel studied the patterns of inheritance by observing seven different features of garden peas. Each of these seven features existed in one or another of two alternative states. For example, pods could be yellow or green, and seeds could be round or wrinkled. How many phenotypically distinctive kinds of pea plants could Mendel have found? Answer: 128. Two kinds of flower color, times two kinds of flower location, times two kinds of plant height, times two kinds of seed color, times two kinds of seed shape, times two kinds of pod color, times two kinds of pod shape.

$$2 \times 2 \times 2 \times 2 \times 2 \times 2 \times 2 = 128 = 2^7$$

In other words, if one gene exists in 2 allelic forms with dominance in effect, then 2 phenotypes are possible. If 2 genes exist and each has 2 allelic forms, then 4 different phenotypes are possible. If 3 different genes exist and each has 2 allelic forms, then 8 different phenotypes are possible. If 4 different genes exist and each has 2 allelic forms, then 16 different phenotypes are possible.

A pattern is emerging that can be expressed mathematically: number of possible phenotypes = X^y, where X is the number of different alleles per gene and y is the number of different genes. Let's assume for the sake of illustration that there are 300 different genes involved in producing the kinds of human characteristics that were mentioned earlier. Let us assume further that each of those genes has only 2 allelic forms. The number of different phenotypic combinations possible (if we assume dominance is in effect for each gene) is 2^{300}, or 2.0369×10^{90}. It is extremely difficult to imagine the enormity of that number. At the moment there are approximately 5 billion people inhabiting the earth, or 5×10^9 people. If we assume, for ease of calculation, that a generation lasts 25 years and thus every 25 years there are a new 5 billion people, and that this has been happening for the last million years or 40,000 generations, then in the last million years the earth has seen 2.0×10^{14} different people. In other words, in the last million years there have not been anywhere near enough people produced to exhaust the genetic diversity of only 300 genes. In reality, human beings probably have thousands of genes and many of these genes can have more than 2 alleles.

ARE ALL ALLELES EQUALLY GOOD?

One hidden assumption in the preceding mathematical model is that it makes no real difference to the organism which of two alternative phenotypes are expressed. In much of the discussion of Mendelian genetics very little relevance to the functioning of the organism was attached to the presence or absence of a particular allele. What difference could it make to a pea plant if it produced red flowers or white flowers? In some cases there does not appear to be any particular advantage or disadvantage associated with the presence or absence of alternative alleles. However, in many other cases—as, for example, in the gene responsible for the production of the beta chain of human hemoglobin— the presence of one allele or the other makes a significant difference to the functioning of the individual.

Individuals who are homozygous for the sickle-cell allele suffer from all the difficulties described in Chapter 6 and suffer a shortened life span. People who die at a relatively young age are less likely to produce offspring than are those who live longer. Early death not only removes the individual from the population but removes his or her genes as well. If proportionately more people with sickle-cell anemia die early than do people carrying the normal allele, then the occurrence of the sickle-cell allele in the population will decrease. The less common the allele is in the population, the less often the corresponding phenotype will appear in the population. The disproportionate death rate of people homozygous for sickle-cell anemia can result in a change in the phenotypic composition of the population.

CHARLES DARWIN

One of the first persons to achieve a clear understanding of the significance of individual variation was a nineteenth-century Englishman named Charles Darwin. Darwin had served as the naturalist on a cruise around the world aboard H.M.S. *Beagle* and was particularly impressed by two things. Not only were there great differences among the different species that he encountered, but there were also many species that shared a large number of similarities. Of particular interest to Darwin was an assemblage of finches found on a group of islands called the Galapagos (see Figure 7.4). Each island in the group had its own species of finches, which differed principally in the size and shape of the beak. The beaks, of course, are the main food-gathering organs of birds and allow each species of bird to specialize on a different source of food

Figure 7.4

The fourteen species of finches inhabiting the Galapagos Islands show strong similarities to each other. This high level of similarity has been interpreted as evidence that they share a common ancestry. [© David Lack, *Darwin's Finches,* Cambridge, England, Cambridge University Press, p. 19, 1961.]

materials. Darwin observed that while each species of finch differed from the others, all were exceedingly similar to one another and to other species of finches on the distant South American mainland. It appeared that some basic ancestral finch type had been modified to occupy each of the several different islands.

Darwin pondered on his observations for years before publishing his ideas in the now famous *On the Origin of Species.* Actually, Darwin would not have published the *Origin* in 1859 as he did had it not been for a letter from Alfred Russell Wallace. This young naturalist, working in Southeast Asia, had reached essentially the same conclusions as Darwin, and he wrote to Darwin to ask what he thought of the idea. This communication resulted in both men presenting their ideas jointly at a scientific meeting in London in 1858. The train of reasoning that both men followed was based on four generalizations derived from observational experience.

Individual Variability In a population of organisms no two individuals are ever exactly alike. Individual variability seems to be a law of nature. This is a simple restatement of the kind of observation that all of us have made and probably requires no elaboration.

Heredity Both men inferred that there had to be some kind of strong hereditary principle to account for the observation that there is a strong tendency for offspring to most closely resemble their parents. Again this is a restatement of a very simple observation. While all individuals are different, there are degrees of difference, and close relatives tend to show a much lower degree of difference than do those who are more distantly related. This statement can be put in another way by saying that those distinctive features possessed by an individual are more apt to appear in that individual's offspring than to appear in the offspring of another. In human terms this means that you are more likely to resemble your mother than to resemble some unrelated lady who lives in Hoboken, New Jersey. While Darwin and Wallace were aware of the necessity of some hereditary principle, neither of them ever discovered what it was.

Biotic Potential Organisms have an incredible ability to produce offspring. The reproductive potential of almost any organism is phenomenal. Everyone has heard about the fabled ability of rabbits to produce more rabbits. Anyone who has tried keeping a female cat as a pet knows that she can produce kittens faster than you can find friends to give them to. Every once in a while you can find a space filler in the newspaper that tells you that if all the offspring from one pair of houseflies were to survive and reproduce repeatedly, at the end of one year the entire surface of the earth would be covered with houseflies to a depth of 40 feet. Female oysters release a million eggs at a time. Even a pair of reproductively conservative elephants could produce enough individuals to form a parade to the moon and back in a scant few thousand years.

Population Stability   Over reasonable periods of time population sizes tend to remain relatively stable. If you count the number of individuals in a population year after year, the numbers do not tend to vary drastically. This population size stability exists despite the reproductive ability mentioned earlier. The inference is clear that death removes individuals from the population as rapidly as reproduction adds them.

The Darwinian Generalizations   Using these pieces of information, Darwin concluded that some of the genetically determined individual differences could add to the ability of those organisms that possessed them to survive and reproduce. Thus the characteristics that improved the chances of survival and reproduction would be likely to appear in the next generation, while those characteristics that did not enhance survival and reproduction would be less likely to appear in the next generation. If this process were repeated consistently, generation after generation, the population would be modified in such a way as to most efficiently utilize the prevailing conditions. The population should achieve a nearly ideal adaptation to the environmental conditions in which it occurred. It is also important to notice that the characteristics of the population of organisms would change with the passage of time.

Darwin's proposal caused an uproar. He claimed that selection by the natural conditions of the existing world (which, understandably enough, became known as natural selection) could bring about the same kinds of changes in wild populations of organisms that people had produced through selective breeding in domesticated plants and animals (see Figure 7.5). Furthermore, this same mindless process could bring new species into existence.

Many biologists of the day accepted Darwin's hypothesis immediately. Their reaction was, "It's obvious; why didn't I think of that?" Others, particularly nonscientists but some eminent biologists as well, objected strenuously. Public discussions, which shed great heat but little light, were held repeatedly. Continuations of these same discussions can still be heard in many places today.

The objections to this hypothesis were many and varied, but in their simplest form they fell into two broad categories. First, Darwin's hypothesis contradicted the biblical account of biological origins as expressed in Genesis, that is, that the Creator has made everything just the way it is. Second, no one had ever seen a species evolve or change into something else. There are, in fact, two entirely different questions being raised here. Do species change? If species do change, what is the mechanism by which this is accomplished?

VERIFICATION OF DARWIN'S THEORY

The first question is much easier to answer in the affirmative today than it was a hundred years ago. As an example, take the peppered moth of England. Peppered moths have always been relatively common in England, spending the day resting on vertical surfaces such as tree trunks and flying around at twilight. In the early 1800s peppered moths were characterized as dull, grey-tan moths with black and white specks. There also existed a rare, mutant form of the peppered moth that was entirely black. It has been learned subsequently that the black form is controlled by a simple Mendelian dominant allele. It is important to understand that the black form was extremely rare; its frequency was approximately equal to the rate of appearance of black alleles by mutation. Maybe one peppered moth in 50,000 was black. There have been enough butterfly collectors in England for a long enough time to know that up until the middle of the nineteenth century the black form of the peppered moth was extremely rare. Today, except in the southwestern corner of England, the frequency of the two color patterns of the peppered moth is almost completely reversed. It is quite obvious that the standard appearance of the peppered moth in England has changed drastically in the past hundred years. Species do change.

The second question regarding the mechanism by which species change may indeed be open to discussion. However, it is worth noting that Darwin's hypothesis gives the most satisfactory explanation of the events without resorting to untestable factors. The explanatory power of Darwin's generalization has been so consistently successful that it is commonly referred to as the theory of evolution by means of natural selection. Once again the case of the peppered moths serves as an excellent example.

The standard-colored peppered moths spend the daylight hours resting quietly on the trunks of trees. These trees frequently have crusty-looking lichens growing on them. A moth at rest among the lichens blends in very well and is easily overlooked by passing humans or birds. It is important for a moth to be so **cryptically** colored because birds eat peppered moths. Black-colored peppered moths, resting on the same surface, are very conspicuous, readily perceived by birds, and rapidly eaten. Needless to say, moths that have been eaten have a lowered probability of reproductive success. Black moths are much

Figure 7.5

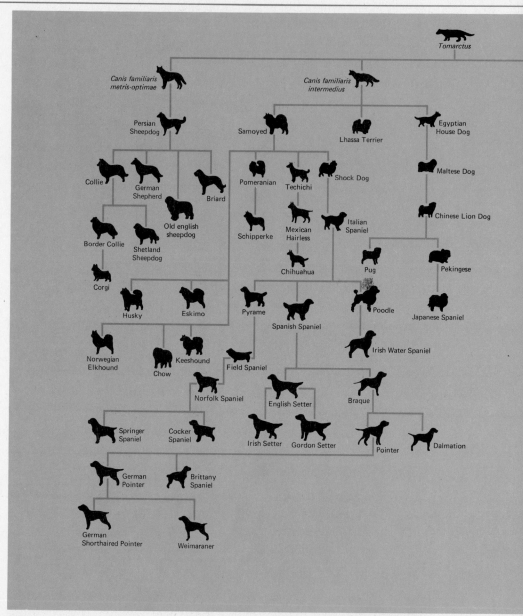

The many breeds of dogs have been achieved by selectively breeding for those features considered desirable. This geneology depicts the ancestral history of 110 separate breeds. [Reprinted with permission of Macmillan Publishing Co., Inc., from *Race and Races,* second edition, by Richard A. Goldsby. Copyright © 1977 by Richard A. Goldsby.]

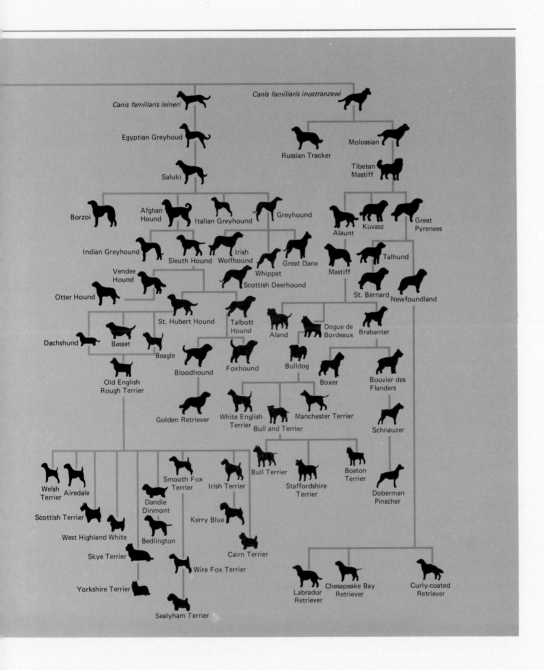

Canis familiaris leineri

Canis familiaris inostranzewi

Egyptian Greyhoud

Russian Tracker

Molossian

Saluki

Tibetan Mastiff

Borzoi

Afghan Hound

Italian Greyhound

Greyhound

Alaunt

Kuvasz

Great Pyrenees

Indian Greyhound

Sleuth Hound

Irish Wolfhound

Great Dane

Talhund

Vendee Hound

Whippet

Mastiff

Otter Hound

Scottish Deerhound

St. Bernard

Newfoundland

St. Hubert Hound

Talbott Hound

Aland

Dogue de Bordeaux

Brabanter

Dachshund

Basset

Beagle

Bloodhound

Foxhound

Bulldog

Boxer

Bouvier des Flanders

Old English Rough Terrier

Golden Retriever

White English Terrier

Manchester Terrier

Bull and Terrier

Schnauzer

Welsh Terrier

Airedale

Smooth Fox Terrier

Irish Terrier

Bull Terrier

Staffordshire Terrier

Boston Terrier

Doberman Pinscher

Scottish Terrier

Dandie Dinmont

Kerry Blue

West Highland White

Bedlington

Skye Terrier

Cairn Terrier

Yorkshire Terrier

Wire Fox Terrier

Labrador Retriever

Chesapeake Bay Retriever

Curly-coated Retriever

Sealyham Terrier

Figure 7.6

(a)   (b)

Biston betularia, the peppered moth, and its black form *carbonaria.* (a) At rest on li-
chened tree trunk in unpolluted countryside; (b) at rest on soot-covered oak trunk near
Birmingham, England. [From the experiments of Dr. H. B. D. Kettlewell, University of
Oxford.]

more commonly eaten because they are much easier to find. Consequently, the black alleles remain very rare under these circumstances
[see Figure 7.6(a)].

Suddenly the industrial revolution comes to the midlands of England. Cities such as Manchester and Liverpool acquire thousands of
coal-burning factories that emit tons of smoke and soot. The soot has a
way of settling out of the air and coming to rest on the rocks, ground,
buildings, and trees, turning everything to a lovely shade of black. The
soot also kills lichens. Normal-colored peppered moths, resting against
a background of blackened tree trunks, are no longer cryptically colored. They are conspicuous and hence are rapidly eaten by birds. Black
peppered moths are not at all conspicuous, are more often overlooked
by birds, and are much more likely to reproduce [see Figure 7.6(b)]. The
next generation of peppered moths is going to have a much higher incidence of black moths than preceding generations because the moths
possessing the black alleles have been able to reproduce. Changed environmental conditions in the midlands of England have enabled the

appearance and the genetic composition of peppered moth populations to change markedly.

But what about the peppered moths living in southwestern England? As was mentioned earlier, the peppered moths in this part of the country look very much as they did a hundred or more years ago. Why haven't these peppered moths turned black as well? The answer is simply that the southwestern corner of England never became heavily industrialized. Large numbers of factories did not get built, large volumes of smoke were not produced, and the environment was not blackened. Hence the black form of the peppered moth is still at a disadvantage in rural England and the normal-colored form still prevails. Natural selection may work in different ways in different places or at different times.

It is important to stress two aspects of the theory of evolution by natural selection that are frequently overlooked. The first of these is that evolution is a phenomenon of populations. It is the allelic composition of a population that is changed from one generation to another. Individuals may change in their lifetimes, but these changes are developmental and do not involve any alteration of the genetic composition of the individual. Hence we must conclude that individuals cannot evolve; only populations can. The second point worth noting is that the selective value of a particular allele is determined by the environmental context in which it must function. There is nothing intrinsically good or bad (better or worse) about one allele or its alternative. It all depends upon those conditions surrounding the workings of the allele. Notice that either of the two alleles for color in the peppered moth can be favorable or unfavorable, depending upon the color of the background against which the moths rest.

This concept of relative goodness of alleles can be emphasized by a reconsideration of the example of sickle-cell anemia. From what has been described so far, almost everyone would be willing to classify the sickle-cell allele as a bad allele, one that is deleterious to its possessor. Imagine the surprise of geneticists and evolutionists when they found human populations in eastern Africa where this bad allele occurred at a frequency of 20 percent. (In a population of 50 people, each with two genes for the beta chain of adult hemoglobin, there would be 20 alleles for sickle-cell hemoglobin.) This is an unexpectedly high incidence for an allele that causes all the kinds of problems that have been described earlier.

The geographic areas that have a high frequency of sickle-cell alleles are also areas in which the occurrence of a form of **malaria** has been high historically (see Figure 7.7). Malaria is a disease of red blood cells caused by a single-celled organism transmitted by mosquitoes. Malaria is a severely debilitating disease that is often fatal. Normal red

Figure 7.7

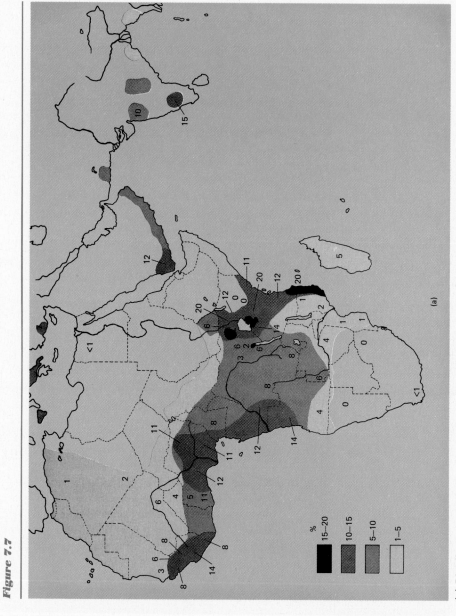

(a) Distribution and frequency of the sickle-cell allele occurring in people of the Old World.

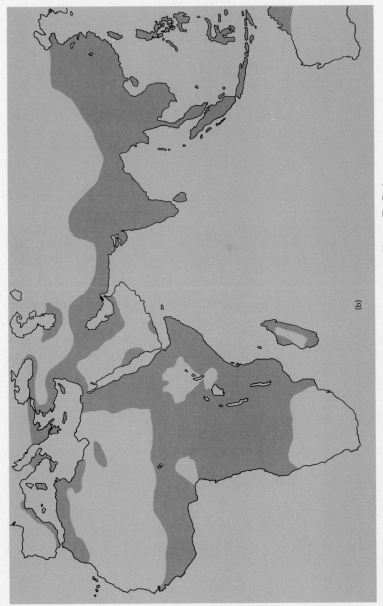

(b) Distribution of a form of malaria in the Old World tropics. [From T. Dobzhanski, *Mankind Evolving.* Copyright © Yale University Press.]

blood cells and normal hemoglobin are prime targets for the malarial microorganism. Sickled red blood cells or sickled hemoglobin are more resistant to malarial parasites. In East Africa a person homozygous for the sickle-cell alleles has the anemia and a decreased life expectancy. A person homozygous for the normal alleles has normal hemoglobin and runs a high risk of contracting malaria, thus having a decreased life expectancy. A person heterozygous for both alleles has some lowered oxygen-carrying capacity in his or her red blood cells but also has greatly increased resistance to malarial parasites. The heterozygotes have a greater life expectancy than either of the two homozygotes. The cost of maintaining a high proportion of heterozygotes in the population is the production of a large number of homozygotes with short life expectancies.

In a case such as this it should be apparent that neither allele is good or bad because both must be present in order to maximize the chances of human survival. The sickle-cell allele evolved as a mechanism for dealing with a significant environmental stress, malaria, and it is only when the allele is removed from a malarial environment that its deleterious character becomes apparent.

HARDY AND WEINBERG

One of the few true scientific laws to be found in biology is called the Hardy-Weinberg law, and it is one of the most puzzling, difficult laws for beginning students to comprehend. Part of this difficulty arises from the fact that it is expressed in mathematical terms and part arises from its appearance of being expressed backwards. What G. H. Hardy, an English mathematician, and H. Weinberg, a German physician, independently discovered around 1908 was a mathematical proof that allelic frequencies could not change and hence populations could not evolve, as long as certain assumptions were met. In reality, these assumptions are not met, so allelic frequencies do change, and populations must evolve.

Take a population of organisms in which a gene exists as either one of two alleles, A and a. In the total population of genes some will exist as A and the rest will exist as a. Let P equal the **proportion** of genes found as the A allele and let Q equal the proportion of genes found as the a allele. This can also be stated as P equals the **frequency** of the A allele while Q equals the frequency of the a allele. If A and a are the only two alleles for the gene, then P plus Q equals 1. That proportion of the alleles that are A together with that proportion of the alleles that are a must constitute all the alleles in the population.

Assume that in the process of gametogenesis the frequency of alleles in the gametes produced is the same as the frequency of alleles in the diploid generation. That is, if the frequency of A in the population is P, then the frequency of male gametes bearing A is P, and the frequency of female gametes bearing A is also P. Of necessity this means that the frequency of male and female gametes bearing a is Q. Assume further that the population of gametes is infinitely large and that the only factors affecting the ability of male and female gametes to combine is the relative abundance of each of the two alleles.

One would expect the frequency of diploid genotypes in the zygotes produced to equal the products of the frequencies of the involved alleles. For example, the frequency of zygotes with the genotype AA should equal the probability of a male gamete bearing A fertilizing a female gamete bearing A. As the frequency of each of these is P, then the frequency of AA zygotes should be P times P, or P^2. Similarly, the frequency of zygotes with the genotype aa should equal Q^2. The frequency of heterozygous zygotes, Aa and aA, will equal the frequency of A-bearing male gametes fertilizing a-bearing female gametes and the frequency of a-bearing male gametes fertilizing A-bearing female gametes, or P times Q plus Q times P, or $2PQ$. As you may have noticed, this is a biological example of the **binomial expansion** formula. If P plus Q equals 1, then $(P$ plus $Q)^2$ equals 1^2, which is P^2 plus $2PQ$ plus Q^2 equals 1. In these formulas P equals the frequency of A and Q equals the frequency of a, while P^2 equals the frequency of AA, $2PQ$ equals the frequency of Aa, and Q^2 equals the frequency of aa. The following diagram illustrates these ideas.

	$P(A)$	$Q(a)$
$P(A)$	$P^2(AA)$	$PQ(Aa)$
$Q(a)$	$PQ(aA)$	$Q^2(aa)$

$P^2(AA) + 2PQ(Aa) + Q^2(aa) = 1$ (all the genotypes)

What is startling about this mathematical game playing is that the frequencies of alleles in the new diploid generation of the population are exactly the same as the frequencies in the preceding diploid generation. For example, in the new generation of zygotes the frequency of a in the population is now

$$\frac{2Q^2 + 2PQ}{2(P + Q)} = \frac{Q^2 + PQ}{P + Q} = \frac{Q(Q + P)}{(P + Q)} = Q$$

This discussion is best clarified with a concrete example. Let P equal the frequency of A and let Q equal the frequency of a. P plus Q must equal 1. Suppose in a given population P equals 0.7. Then Q must equal 0.3.

	0.7A	0.3a
0.7A	0.49AA	0.21aA
0.3a	0.21Aa	0.09aa

The frequency of diploid genotypes in this new generation of the population is 0.49AA, 0.42Aa, and 0.09aa.

In a population of 200 genes from this population, there would be $2(49) + 42A$ alleles. Thus the frequency of A in the new generation of the population is $\frac{140}{200}$, which is 0.7. The frequency of a is [0.42 plus $2(0.09)$]/2, which is 0.3.

Regardless of how many times the population is allowed to reproduce and randomly recombine alleles, the allelic frequencies remain constant or achieve **genetic equilibrium.** If genetic equilibrium is achieved, if the allelic frequencies remain constant, it means that the population cannot evolve.

There are several hidden assumptions in the Hardy-Weinberg law that allow it to work mathematically. The first of these is that the population must be infinitely large so that sampling errors will have minimum effects in altering the expected recombination of gametes. Second, alleles are not allowed to mutate. Third, no allele may get preferential treatment with respect to remaining within or leaving the population. Thus both alleles must function equally well within the population, and neither can affect a tendency to leave or join the population. Fourth, and finally, there can be no tendency for assortative mating; genotypes must be able to mate without respect to their genotypes. The only thing that is allowed to influence the recombination of gametes is the relative abundance of the two alleles. If any of these assumptions are not met, then genetic equilibrium cannot be maintained. In nature, one or more of these assumptions is usually not fulfilled.

Few populations are infinitely large, although occasionally some appear to be large enough to effectively approximate infinity. More often the effective size of a breeding population is quite small, introducing the possibility of changes in allelic frequencies due to chance sampling errors. As an example, imagine two populations in which the recessive phenotype accounts for 10 percent of the individual organisms. One of the populations consists of 10 individuals and the other consists of 1000. Suppose, by pure chance, one homozygous recessive individual dies in each of the two populations. In both cases 2 recessive alleles are removed from the population. In the larger population this removal of 2 recessive alleles has almost no effect on the frequency of recessive alleles in the population. (If $Q^2 = 0.10$, then $Q = 0.3162$. In a population of 2000 genes, 632 are recessive. If 2 recessive alleles are removed, that leaves 630 in a total population of 1998, or $Q = 0.3151$.) However, the removal of 2 recessive alleles from a population whose total size is 20 will cause a major shift in allelic frequencies. (In a population of 20 genes, 6 are recessive. If 2 of these are removed, that leaves 4 in a population of 18, meaning Q now equals 0.2222.) Small population size can disrupt genetic equilibrium.

As we have already seen, genes do occasionally mutate. An allele can be modified so that it becomes another allele. Anytime this happens, it will change the values of both P and Q. If both P and Q are subject to change by mutation, it is obvious that genetic equilibrium cannot be achieved.

We have also noted previously that different alleles do not necessarily function equally well in the same environments. Those alleles that detract from the organism's functioning tend to be removed from the population, and, conversely, those that enhance survival and reproduction tend to be disproportionately passed on to the next generation. These tendencies will disrupt genetic equilibrium.

It seems quite apparent that in many sexually reproducing organisms, mating is not a completely random mixing of gametes. Some genetically determined characteristics affect mating itself. For example, there is a great deal of individual variability in size among Canada geese. Males are generally larger than females, but there is a great amount of overlap in size. An examination of individual mated pairs of these birds will almost always reveal that the male is larger than the female (see Figure 7.8). It seems reasonable to assume that males prefer females smaller than themselves, or that females prefer males larger than themselves. To the extent that size is genetically determined, this will prevent the establishment of genetic equilibrium.

If all the assumptions behind the Hardy-Weinberg law are false, and the conclusion based on those assumptions is demonstrably incorrect, then why is it such an important biological generalization? The real value of the Hardy-Weinberg law is that it serves as a model for what must happen if the assumptions are met. When we find that genetic equilibrium does not prevail, then we are forced to examine the

Figure 7.8

Many species of organisms do not mate at random. As exemplified by this pair of Canada Geese, a female goose prefers a male that is significantly larger than herself. The collars on the necks of these birds are an artificial means of individual identification for long-term field studies. [Photo courtesy of Dr. B. C. Lieff.]

population more carefully to determine which assumptions in the law are being violated. It is largely the violation of Hardy's and Weinberg's assumptions that provides the mechanisms for validating Darwin's theory.

Speciation

As originally defined in Chapter 2, a species consists of all those individual organisms that are capable, or potentially capable, of interbreeding. The addition of *potential capability* in the definition allows us to include within the species those individuals that die before being able to reproduce, those that are sick, those that are isolated and physically prevented from reproducing, and so forth. Implicit within the definition of a species is the notion that individuals of one species do not interbreed with individuals of some other species. There is something in the nature of the species, something intrinsic to the species itself, that prevents it from interbreeding with another species.

If a species consists of individual organisms, all of which are capable of interbreeding — that is, all of which are capable of exchanging genes — then it should also be possible to think of a species as an abstract pool of genes that can be separated and recombined generation after generation. A species consists of all those individuals who are a part of the same **gene pool.** Because we have seen that the allelic composition of this gene pool may change through time, a species may be conceptualized as an independently evolving gene pool or an independently evolving lineage. Because the genes of one species do not mix with the gene pool of a second species, the two lineages must evolve independently.

When one examines the vast diversity of species found inhabiting the world today, one quickly notices that some independently evolving lineages look much more alike than do others. For example, white pine trees look nothing like upland chorus frogs, but scrub pines look a lot like white pines and spring peepers look a great deal like upland chorus frogs (see Figure 7.9). The number of parallel examples of this phenomenon is huge. This observation — that all species are different but some are more different than others — is very similar to the observation made earlier that all individuals are different, but some are more different than others. Within a species we have noted a strong tendency for offspring to most closely resemble their parents, their immediate ancestors.

Is it possible to use the same reasoning to explain the similarities that we find between species? Could the high level of similarity found between two independently evolving lineages, such as the spring peeper and the upland chorus frog or the white pine and the scrub pine, be due to the fact that they have some ancestor in common? It is certainly a question worth investigating.

This possibility of two similar-appearing species having a common ancestry with one previously existing species was (and for some

Figure 7.9

(a)   (b)

(c)   (d)

Some species look more alike than others. (a) Spring peeper; (b) upland chorus frog; (c) white pine; (d) Virginia pine. [Photos (a) and (b) courtesy of John E. Wiley. Photos (c) and (d) courtesy of Matthew E. Stockard.]

people still is) a major objection to the acceptance of Darwin's theory. It's hard enough to accept the idea that a species is changeable through time, but that it can split into two or more separate species may be too much. If a species is defined as an interbreeding population, reproductively isolated from all other such interbreeding populations and thus forming an independently evolving lineage, how can this one population be converted into two or more populations, each of which will function as an independently evolving lineage?

GEOGRAPHIC ISOLATION MODEL

The easiest way for replication of independently evolving lineages to be accomplished is to have the ancestral population geographically separated by some **extrinsic** barrier. An extrinsic barrier is defined as any external factor that prevents gene exchange between one part of the population and another. In its simplest form it is an area that is uninhabitable and essentially uncrossable by members of the ancestral population. For example, to a forest-living organism an extensive area of grasslands could function as a barrier, and to a grassland-dwelling organism an extensive area of forest could function as a barrier. A mountain range could be a barrier to lowland forms, and vice versa. Oceans can prevent gene flow between terrestrial populations, and continental land masses can prevent gene flow between marine populations.

Once a barrier has become established, separating a population into two (or more) disjunct subpopulations, then the subpopulations must evolve independently, because according to the definition of a barrier they are incapable of exchanging genes. Two different factors immediately go into effect. The first of these is that the physical conditions on either side of the barrier will most likely be different. Thus the subpopulations on either side of the barrier will be existing under different environmental regimes and will most probably adapt to those different conditions. Secondly, new mutations, chromosomal rearrangements, crossovers, recombinations, and so forth, which all arise randomly, have no means of crossing the barrier from one subpopulation to the other. Thus the genetic diversity present in each of the two subpopulations will also differ.

As more and more time passes, the genetic composition of the two subpopulations must become more and more different because of the independent arising of genetic diversity and the separate selective conditions. As long as the barrier remains in place, preventing the two subpopulations from exchanging genes, they will continue to grow more distinct. They will diverge. They will evolve independently. This general pattern is shown in Figure 7.10.

At what point do geographically isolated, genetically different populations become species instead of being subpopulations of the same species? This question is nearly impossible to answer, because the definition of *species* hinges on the potential for interbreeding within a species and the lack of interbreeding between species. The

Figure 7.10

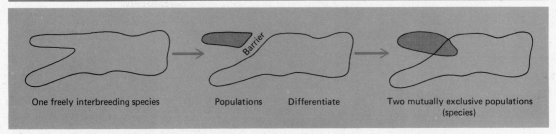

One freely interbreeding species   Populations   Differentiate   Two mutually exclusive populations
                                                                                                    (species)

Geographic isolation model of speciation.

presence of the extrinsic barrier makes it difficult, if not impossible, to determine the intrinsic limits to interbreeding. The only certain way of determining species status is to have both populations naturally occur in the same place at the same time and not interbreed.

It is important to stress the importance of normal nonbreeding between species. Getting two individuals from disjunct populations to breed successfully in captivity proves very little other than a certain level of genetic similarity. Let me illustrate this contention with an example. Lions and tigers have both been bred in captivity, and on rare occasions it has been possible to obtain **hybrid** offspring between lions and tigers. These strange cats have been called ligers or tiglons (see Figure 7.11). Despite this evidence of interbreeding, lions and tigers are still regarded as two separate species. The reason for this is simple. While lions are essentially animals of the African plains and tigers are essentially animals of the Asian forests, there is an area in western India where their geographic ranges overlap. In this area lions breed exclusively with lions and tigers breed exclusively with tigers, and the two forms of cats ignore one another as potential mates. The western Indian populations of lions and tigers conclusively demonstrate that these are two different species of organisms. The critical feature in **speciation** is the establishment of intrinsic mechanisms that prevent interbreeding.

The acid test of speciation occurs when the extrinsic barrier is removed and the two populations reestablish contact. If both subpopulations freely interbreed, producing vigorous, viable intermediate offspring that are capable of interbreeding with each other and either parental form, then it is apparent that no intrinsic isolating mechanisms have developed, and we are still dealing with one very variable species. However, if members of each subpopulation breed only with members of their own subpopulation and actively ignore members of

Figure 7.11

This tiglon, offspring of a male tiger and a female lion, resided at the Central Park Zoo in New York from 1938 until 1953. While rare hybrids of this kind have been brought about in captivity, they are not known to occur in the wild. [Photo courtesy of The New York Times.]

the other subpopulation as potential mates, then it is equally apparent that intrinsic isolating mechanisms have developed during the period of separation, and we are now dealing with two distinct species.

While these two potential outcomes are clearly different, it is possible to imagine many degrees of intermediate situations when the two populations make initial contact. Eventually one of two circumstances has to prevail. Either the hybrid offspring (those produced by unlike parents) function as well as or better than the offspring of like parents, or they function less well. If the former circumstance holds true, then all individuals are capable of freely exchanging genes, all are part of the same gene pool, and we still have one species. If the hybrids do not function as well as the offspring of like matings, then it is apparent that fewer of the hybrids will contribute to the formation of subsequent generations. Furthermore, those parental types with faulty powers of discrimination that result in the production of hybrid offspring are also selected against to the extent that hybrid offspring do not do as well as like-parent offspring. To put it the other way around, those adults least likely to mate with adults of the other population are most likely to produce the maximum number of surviving and reproducing offspring, and those factors that minimize hybridization are those most likely to be passed onto subsequent generations. This situation should result in even greater differentiation between the two populations, making their status as distinct species certain.

It is generally believed that this model of speciation induced by geographic isolation is the most common explanation for the observed level of diversity among living populations. However, at least two other mechanisms have been proposed, which may have some validity under some circumstances.

SPECIATION BY POLYPLOIDY

In the normal course of the reproductive events outlined, haploid gametes of the same species fuse to produce diploid zygotes, and at some subsequent time meiosis produces more haploid gametes. All this is based on the normal functioning of mitosis and meiosis. Sometimes— more often in plants than in animals—the mitotic process goes a little haywire, and while all the chromosomes are replicated, cytokinesis does not occur. Thus the resulting cell has four of each kind of chromosome, a condition called **tetraploid.** If the tetraploid cell then undergoes normal mitosis, all cells derived from it will also be tetraploid. If any of these cells become gonads, which undergo normal meiosis, all the gametes produced will be diploid. If diploid gametes fuse, they produce tetraploid zygotes, and a new species has been produced. Tetraploid plants are seldom capable of interbreeding with their diploid ancestors, so this kind of **autopolyploidy** is an example of instant speciation (see Figure 7.12).

Again more often in plants, gametes from two different species sometimes fuse to produce hybrid zygotes. Under such circumstances the genetic constitutions of the two gametes are so incompatible that the zygote usually dies very quickly. However, sometimes, in the zygote stage, the same kind of disruption of cytokinesis occurs, leaving the cell with two copies of each of the two kinds of chromosomes, a situation known as **allopolyploidy.** This condition is easiest to perceive if the two parents of the hybrid have different chromosome numbers. Subsequent mitotic replications may be completely normal, resulting in a plant with a complete complement of genes from each of the two parents. If this plant can interbreed with another—or, commonly, if it is self-fertile—it is the basis for a new independently evolving lineage. Allopolyploidy can give rise to new species (see Figure 7.13).

SYMPATRIC SPECIATION

Repeated attempts have been made to propose a mechanism whereby species may arise without the necessity of a period of geographic isolation and without the intervention of polyploidy. Because both species are produced in the same geographic area, this process is called **sympatric** speciation. Such proposals all suffer the problem of minimizing gene flow within an interbreeding population. The most likely mecha-

Figure 7.12

(a)   (b)

(c)   (d)

By treating certain plant materials with the chemical colchicine, one can induce a doubling of the normal complement of chromosomes. In each of these forms of day lilies, the figure on the left has two of each kind of chromosome (diploid), while the right-hand partner has four of each kind of chromosome (tetraploid). [From Dr. Toru Arisumi, *Journal of Heredity,* vol. 55, p. 254, 1964. Copyright 1964 by the American Genetic Association.]

nism by which this can be accomplished is through something called *disruptive selection*. Disruptive selection is that set of circumstances in which two different phenotypes are favored but any intermediate phenotypes are disadvantageous. For example, suppose it were advantageous to be either tall or short but selectively disadvantageous to be of intermediate height. Disruptive selection would tend to favor those matings that produced offspring that were either tall or short but not intermediate in size. If these conditions existed long enough, eventually only like phenotypes would interbreed, producing only the favored phenotypes of offspring. In effect, two independently evolving lineages would be produced. Whether or not this process actually occurs in nature, and its relative importance as a mechanism of specia-

Figure 7.13

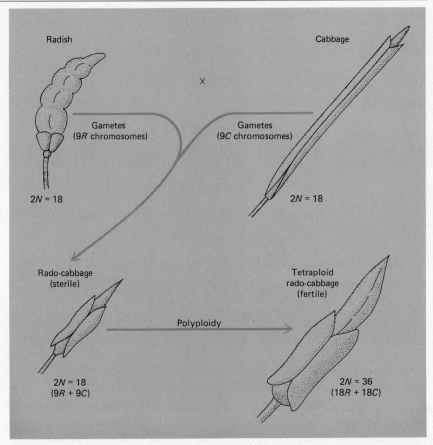

A diagrammatic presentation of the fruits of radish, cabbage, and their hybrid off-spring. The diploid hybrid (with 18 chromosomes) is not capable of reproducing, but the allotetraploid (with 36 chromosomes) is completely fertile. Tetraploid rado-cab-bages are bigger than either parent, with the roots like a cabbage and the leaves like a radish. [After Karpechenko.]

tion, is still being investigated. It has been suggested that this proposal may be of great significance in understanding the origin of species that differ principally in their choice of major food items, such as many plant-eating insects.

Diversity and Systematics

The concepts of evolution and speciation provide us with a new per-spective for examining the tremendous array of diversity presented in living forms today. If all of the species in existence today are derived from previously existing forms, then it is most logical to assume that those species that are most alike are those that most recently shared a common ancestor, and those that are less alike share a less recent common ancestry. The high degree of similarity found between some species is presumably due to the retention of a large number of genes in common with those of the common ancestor.

By convention, those species that have a great number of characteristics in common and probably share a relatively recent common ancestor are said to be members of the same **genus.** While it is possible to define the species on the basis of interbreeding among the individual organisms within the species, such precision is not possible for a genus or any other more inclusive group. The genus is a rather subjective and arbitrary grouping of similar species. Those species that are not similar enough to be grouped into the same genus are placed in other genera (plural for *genus*). For example, wolves, coyotes, and dogs have a great number of characteristics in common, probably share a relatively recent common ancestor, and are all considered members of the genus *Canis.* Foxes are somewhat similar to the species above but not similar enough to be thought of as having as recent a common ancestor. Therefore, the foxes are placed in other genera (namely, *Vulpes* and *Urocyon*).

Similar genera are also thought to have a common ancestry, although not as recent as the common ancestry of species within a given genus. This relationship among genera is indicated by grouping them together in the same *family.* For example, all the doglike species included in the genera *Canis, Vulpes,* and *Urocyon* are included in the family Canidae, while all the catlike species included in the genera *Felis, Panthera,* and *Acinonyx* are grouped together in the family Felidae. With this same pattern of reasoning, species are grouped in more and more inclusive categories. Different families sharing basic similarities are grouped into an *order.* Orders may be grouped into a *class.* Classes are united to form a *phylum* or *division.* Similar divisions (in the case of plants) or phyla are grouped together to form a *kingdom.* As an illustration, the classification scheme for the lion is shown below.

Kingdom	Animalia
Phylum or Division	Chordata
Class	Mammalia
Order	Carnivora
Family	Felidae
Genus	*Panthera*
Species	*leo*

Such a system of **hierarchical** categories not only allows us to keep track of (**classify**) a huge amount of organic diversity but also gives us some idea of how the array of diverse organisms has come to exist. The rank of a category—the level at which species are grouped together—has very strong implications about the recency of their common ancestor and their degree of relatedness. In effect, this system of classification should allow us to reconstruct the **phylogeny** (evolutionary history) of the species included, at least in a rough way.

Interestingly enough, this system of classification by hierachical categories was proposed by a Swede, Carl von Linné, around the middle of the eighteenth century, a full century before the idea of species' evolution was proposed. Linné's purpose was simply to make a listing of all the organisms in the known world in such a way that information about each species could be found easily. Keeping similar species near each other in his books accomplished this purpose. Linné, assisted by the many people who collected specimens in various parts of the world, achieved a great deal of success and recognition throughout Europe and is known more commonly by the latinized form of his name, Carolus Linnaeus.

Linnaeus's most significant contribution to biology is in the consistent use of **binomial nomenclature.** Before Linnaeus, many people had written books about species of plants and animals but had been inconsistent and cumbersome in referring to particular species. Each species had a different common name in each European language (sometimes more than one), and the learned texts of the day, which were usually written in Latin, referred to each species by a long description. Linnaeus referred to each species by a two-word name. The first word was a noun based on either a Latin or Greek word, and the second word was an adjective (in Latin or Greek) modifying the noun. The first word was the genus or generic name and the second word was the species or specific name. The first letter of the generic name is always capitalized (it is a proper noun), the specific name is not capitalized, and both names are set in italics. (When written, scientific names of species are underlined as the conventional method of informing typesetters to print the words in italics.) Some examples of scientific names are shown in Table 7.1. This method of naming species is universally accepted by biologists today and does much to facilitate communication.

One of the advantages of a system of hierarchical classification is that a great deal of information is communicted by the use of a simple categorical name. Reference to a particular family of species automatically tells you that these species possess all the characteristic features used to describe the kingdom, phylum, class, and order in which this family is included, as well as the common characteristics shared by the constituent genera. For example, a reference to the family Felidae tells you not only that we are talking about catlike genera but that we are referring to animals with backbones and skulls, that are warm-blooded and covered with hair, that are four-legged and carnivorous, and proba-

Table 7.1
Examples of Scientific Names

COMMON NAME	GENUS AND SPECIES
Man	*Homo sapiens*
Dog	*Canis familiaris*
Cat	*Felis catus*
Cow	*Bos taurus*
Horse	*Equus caballus*
American robin	*Turdus migratorius*
European robin	*Erithacus rubecula*
Turkey vulture	*Cathartes aura*
Black racer (snake)	*Coluber constrictor*
Timber rattlesnake	*Crotalus horridus*
Snapping turtle	*Chelydra serpentina*
Southern toad	*Bufo terrestris*
Oak toad	*Bufo quercicus*
Chorus frog	*Pseudacris triseriata*
Atlantic salmon	*Salmo salar*
Largemouth bass	*Micropterus salmoides*
White pine	*Pinus strobis*
White oak	*Quercus alba*
Carrot	*Daucus carota*
Garden pea	*Pisum sativum*

bly have retractile claws on their toes. The name Felidae implies all those characteristics of the order Carnivora, the class Mammalia, the phylum Chordata, and the kingdom Animalia, without having to state them explicitly. The name Felidae also tells you that all the included species share more characteristics in common with one another than any one of them shares with any other species not included in this family. The advantages of such a system in enabling us to discuss the wealth of diversity displayed by living species cannot be overstated.

Obviously, the highest categories of a classification should allow a person to make some reasonable generalizations about all the species included within it. For many years people tried grouping all the species into one or the other of the two most inclusive categories, the kingdoms. Living species had to be considered as either plants or animals. This approach to classification has been abandoned by most biologists because it is not adequate for describing the known diversity among species. The terms *plant* and *animal* convey very complex ideas about the kinds of organisms being discussed. The amount of detail used to elaborate the distinctions between plants and animals does not really alter the meaning of those concepts. The real problem is that there are many species of organisms that are not clearly plantlike or clearly animallike. Many of these species possess characteristics that are part plantlike, part animallike, and some completely unique. Attempting to broaden the original concepts of plants and animals to include these species makes the familiar categories almost meaningless.

Instead, many biologists today are attempting to work with a classification based on five kingdoms. The old familiar kingdoms of plants and animals are clarified and supplemented by the recognition of three additional kingdoms of organisms that don't fit very easily into the two-kingdom system. These new kingdoms include the Monera, the **Protista,** and the **Fungi.**

Monera, as mentioned in Chapter 2, includes all those organisms composed of procaryotic cells. These organisms are most certainly more like one another than any one of them is like any eucaryotic organism. In most respects the differences between the monerans and any eucaryotic organisms are probably greater than the differences between plants and animals. The fungi are mostly multicellular eucaryotic organisms that have specialized in decomposition (a sort of passive heterotrophism) as a way of life. Plants are multicellular eucaryotic organisms that have specialized in photosynthesis as a way of life. Animals are multicellular eucaryotic organisms that have specialized in predation and parasitism (much more active heterotrophism) as their way of life. The kingdom Protista is a diverse assemblage of mostly single-celled eucaryotic organisms that don't conveniently fall into one of the other three eucaryotic kingdoms. They tend to be much more diverse in their modes of existence than are the plants, animals, or fungi. Even in those species that are clearly multicellular, there are strong indications that these species share many more characteristics in common with unicellular protistans than they do with either plants, animals, or fungi.

Summary

1 / *Living things exhibit both interspecific and intraspecific variability.*

2 / *The amount of genetic variability that can be generated by sexual reproduction is astronomically huge.*

3 / *All genotypes may not function equally well in the real world.*

4 / *Charles Darwin and Alfred Russell Wallace proposed that genetic differences that enhanced survival and ability to reproduce would be most apt to predominate in subsequent generations. Repeated modifications of the population in this way could result in descendants being phenotypically different from their ancestors.*

5 / *Darwin also suggested that this process of natural selection could explain the repeated origin of new species, thus accounting for why some species are more alike than others.*

6 / *The change in appearance of populations of peppered moths in the midlands of England over the last century can be most easily explained as the result of natural selection induced by the change in background color brought about by industrialization.*

7 / *Only populations can evolve.*

8 / *The direction in which a population evolves depends upon the conditions in which the population exists and the particular genetic characteristics that appear.*

9 / *Natural selection may work in different ways at different times or places or on different populations.*

10 / *The allele for sickle-cell anemia can be advantageous in certain tropical regions because it confers resistance to the disease malaria.*

11 / *Hardy and Weinberg demonstrated mathematically that populations achieve genetic equilibrium, that is, cannot evolve, unless certain assumptions are violated. In the real world most of these assumptions are not valid.*

12 / *The production of new species is usually brought about by the disruption of interbreeding within a species. This is most likely to occur through the establishment of geographic isolation. Disruption of interbreeding may also be caused by polyploidy or intense disruptive selection.*

13 / *Organic diversity is usually classified by using a system of hierarchical categories. These categories, in order of increasing inclusion, are species, genus, family, order, class, phylum or division, and kingdom.*

14 / *All living species are named by using the Linnaean system of binomial nomenclature. This system has done a great deal to facilitate the communication of biologists around the world.*

15 / *Our present understanding of organic diversity favors the use of a classification scheme with five kingdoms: Plants, Animals, Monera, Protista, and Fungi.*

Questions to Think About

1 / How do you account for the existence of so many different phenotypes among the peoples of the world?

2 / Can you explain why some populations of rats are now immune to the effects of certain kinds of rat poisons that were almost totally lethal 20 years ago?

3 / Why is it that some kinds of bacterial infections are most likely to be acquired in hospitals? Why are these kinds of infections the most difficult to eliminate?

4 / Is there a difference between evolution and speciation?

5 / How do you counter the objection that "Evolution is only a theory"?

6 / Suppose you find a population in which the frequencies of the B and b alleles are 0.8 and 0.2, respectively. If the population has established genetic equilibrium what will be the frequency of genotypes in the next generation? What will be the frequency of phenotypes in the next generation (assume dominance)? What will be the frequency of the two alleles in the next generation?

7 / Suppose you find a population in which 84 percent of the individuals show the dominant phenotype and 16 percent show the recessive phenotype. Has this population established genetic equilibrium? What are the allelic frequencies? What are the genotypic frequencies?

8 / How do you account for the existence of so many different species occurring in the world? Why are some species more alike than others?

Facilitating Gene Exchange

Objectives

The student should be able to:

1 / *describe the male and female reproductive systems of human beings.*

2 / *list which endocrine glands and hormones are associated with regulation of the reproductive systems.*

3 / *describe how a negative feedback system works.*

4 / *describe the similarities and differences between estrous cycles and menstrual cycles.*

5 / *explain how hormones regulate a menstrual cycle.*

6 / *describe briefly the hormonal events that occur subsequent to fertilization and thus regulate the reproductive system during pregnancy.*

7 / *describe (using discretion) the usual methods of conception.*

8 / *describe (again using discretion) the various methods of contraception, giving the advantages and disadvantages of each.*

Introduction

From what was said in Chapter 2, you already know that biologists attempt to study life at many different organizational and structural levels. The knowledge and insights gained from these various studies are most meaningful when they are integrated into a coherent overview of the subject. The more information you have from a greater number of different sources, the greater the chances become that your understanding will be more complete. This basic approach to understanding is as valid for the various subdivisions of biology as it is for the subject of biology as a whole.

In an attempt to understand reproduction, we have examined mitosis and meiosis at the cellular level in Chapter 4. Next we rediscovered the genetic principles at a rather abstract mathematical level in Chapter 5. Then we achieved some understanding of these genetic principles at the molecular level in Chapter 6. The knowledge of what is happening at these lower levels of organization has helped us to understand some of the consequences that occur at the population level in the form of evolution and speciation, which were discussed in Chapter 7. But human beings are multicellular organisms, and most of our perceptions and orientations operate at the organismal level. What has not yet been presented is very much information or knowledge about reproduction at the organismal level, which can be integrated with what we now know about reproduction from these other levels.

It is important to keep in mind that reproduction is only one of many activities that any living organism performs. To maximize the chances of the organism's success, the reproductive activity should be fully coordinated with the performance of all those other activities. It should be apparent that the more complex the organism is and the more different activities that organism performs, the more complex its mechanisms of coordination will be. For complex multicellular organisms, such as human beings, the intricacy of the systems that coordinate and control the performance of living functions is remarkably involved.

The main topic for consideration within this chapter is human reproduction, with strong emphasis on its internal controls. However, the kinds of regulatory mechanisms that will be discussed are not unique to human beings nor to the process of reproduction. The factual contents of this chapter are only illustrations of the control mechanisms that are used by multicellular organisms to coordinate the performance of almost all living activities. While the specific details concerning human reproductive regulation are important and interesting in themselves, it is probably of even greater importance to perceive the general schemes by which activities and functions are controlled.

Some Anatomy

The first requirement for reproduction by a diploid organism is the differentiation of some cells in which meiosis can occur. In human beings this differentiation occurs during the seventh week after conception, long before the individual is actually born. These differentiated structures are organs called gonads—**testes** in the male and **ovaries** in the female.

THE MALE SYSTEM

The growth and development of the testes occurs in the individual's abdominal cavity. Just prior to birth the testes and their associated duct work descend out of the abdominal cavity into a fleshy outpocketing between the legs, called the **scrotum.** In human beings the testes remain in the scrotum permanently, although in other mammals, such as squirrels, the testes descend into the scrotum only during the mating season. The scrotum appears to function as a structure for cooling the testes. The haploid cells that result from meiosis in the testes are unable to survive the relatively high, constant body temperature found in the abdominal cavity. The temperature in the scrotum is several degrees cooler than that of the abdominal cavity, which is sufficient to allow survival of the haploid sperm cells (the male gametes).

Microscopic examination has revealed that the major structural units of the testes are tiny cellular cylinders called **seminiferous tubules.** Some of the cells making up these seminiferous tubules are capable of undergoing meiosis. The haploid cells so produced then undergo several mitotic replications and a period of differentiation and maturation, after which they are called **sperm.** The sperm are moved along the inside of the seminiferous tubules into progressively larger ducts, which eventually extend to the outside of the body at the end of the **penis.** Several glands associated with the tubules add a viscous secretion, called **seminal fluid,** into the ducts, which is used to transport and nourish the haploid sperm. Sperm may spend as much as six weeks being stored in the upper reaches of these ducts. The contraction of involuntary muscles surrounding these ducts forces the expulsion of the sperm and seminal fluid.

The unified duct from the paired testes that travels through the penis is also surrounded by some very spongy-walled blood vessels. Under appropriate kinds of nervous stimulation (which are not always readily apparent), small muscles close off the exits of these blood vessels, thus engorging them with blood. This action causes the penis to increase in size and rigidity. Only under these circumstances is it possible to insert the penis into the female reproductive system and to

Figure 8.1

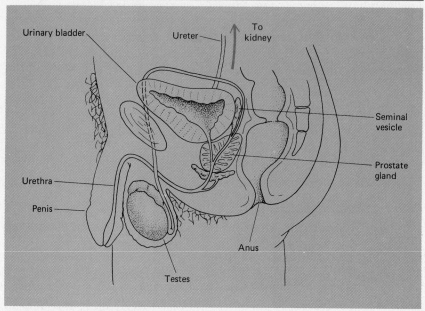

The male reproductive system.

move sperm out of the ducts. The male reproductive system is diagramed in Figure 8.1.

THE FEMALE SYSTEM

The paired ovaries of the female lie in the upper portion of the abdominal cavity, near the kidneys. Each ovary consists of many clusters of tiny fluid-filled receptacles, called **follicles.** It is within the follicle that an egg or ovum is produced by means of meiosis. Nearly surrounding each ovary is a large, membranous, funnel-shaped structure, which narrows to a small, twisted tube called the **oviduct** or **fallopian tube.** The paired oviducts join a larger, more muscular tube called the **uterus,** which opens to the outside of the body by way of the **vagina.** The walls of the vagina are elastic enough to accommodate an inflated penis so that sperm may be transferred from the male system to the female system. Fusion of the gametes, or fertilization, is usually accomplished in the upper third of an oviduct. The female system is diagramed in Figure 8.2.

REGULATING THE SYSTEM

From the description of the structures involved, the act of human reproduction would appear to be relatively simple. That is correct, in the sense that almost any adult person is physically capable of doing so without studying an involved procedures manual. But this apparent simplicity overlooks a great deal of very complex behavioral activity we take for granted. Sights, sounds, scents, and movements all produce responses. These responses are generally in the form of muscular contractions that result in movements of some kind. The functioning of the nervous system is the principal factor determining which muscular responses occur.

It should be apparent that a vast number of different movements are possible, only some of which are appropriate under given circumstances. It requires a very complicated series of interactions between the sensory receptors, the nervous system, and the muscles to ensure that the responses are appropriate to the conditions. But if we ignore all the complexities directly associated with the obvious behavioral as-

Figure 8.2

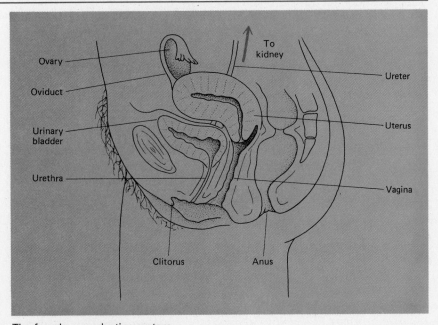

The female reproductive system.

pects of human reproduction, we still find a wealth of complex control systems. These are not very apparent in the males.

Some time during the teenaged years the gonads of the male reach maturity, and the cells of the seminiferous tubules start undergoing meiosis. This process appears to continue, barring some abnormality, until the individual dies. While the rate at which sperm are produced does gradually decline with age, we are dealing with a steady state system, one that runs at an essentially constant rate. The amount of sophistication shown in the regulation of male gamete production is not impressive.

Females, however, demonstrate a totally different pattern of gamete production. Most of us have noticed that female human beings usually achieve sexual maturity two or even three years earlier than males. It is not so obvious but it can also be observed that a woman's release of gametes from the gonads is not the constant, continuous activity that it is in men. Instead, women produce an egg periodically, on the average about once every 28 or 29 days. The release of an egg, or ovum, by the ovary is called **ovulation.** The cyclic nature of ovulation suggests that there must be some rhythmic mechanism at work that controls ovulation. In studying the ovulatory cycle in many different vertebrate animals, biologists have discovered that most of it is regulated by chemical messengers called **hormones.**

Hormones are chemical substances produced by one organ of the body, and they cause a specific reaction in some totally different part or parts of the body. Because the site of a hormone's action is often distant from its site of production, there must be some means of transporting the hormone. This transport is usually accomplished by the circulatory system—in the blood. The organs that produce hormones are collectively known as **endocrine glands,** or ductless glands. Endocrine glands are well supplied with blood vessels, and they secrete their hormones directly into the blood that flows through them. Thus the hormones are distributed all over the body, but only specific organs react to them.

Reproductive Glands and Hormones

There are many different endocrine glands, which produce an even greater number of different hormones. The easiest way to determine which endocrine glands are involved in regulating any function is to compare the appearance and activity of normal organisms with those that lack one specific gland. Any differences are most probably due to the missing gland. If the normal condition can be restored in the glandless organism either by implanting a new gland or by injecting extracts of the missing gland into the organism's bloodstream, then it is even more likely that the gland is responsible for the changes.

Repeated comparative observations of this kind have indicated the involvement of three major endocrine glands in vertebrate reproduction: the **hypothalamus,** the **pituitary,** and the gonads. It is important to stress that not only do the gonads produce gametes, but they also perform a control and regulatory function. The hormonal production of the gonads appears to be regulated by the pituitary, which is in turn regulated by the hypothalamus. In a truly elegant display of circularity, the hypothalamus appears to be regulated by the gonads. A more detailed discussion of these endocrine glands and the hormones they produce should clarify their interrelationships and their effect upon reproduction.

HYPOTHALAMUS

The hypothalamus is a relatively small region at the base of the brain. It is composed of nerve cells that can produce and release hormones. Among the many hormones produced by the hypothalamus is one (possibly more than one) small polypeptide called **gonadotropin-releasing substance,** usually just called GRS. Because of an unusual arrangement of blood vessels, all blood leaving the hypothalamus travels directly to an area between the bottom of the brain and the roof of the mouth. In this area is another endocrine gland called the pituitary.

PITUITARY

Historically the pituitary has been referred to as composed of two parts, the anterior lobe and the posterior lobe. In actuality, the posterior lobe is an extension of the hypothalamus, with its own blood supply, it need not concern us at this point. For the purposes of this discussion the anterior lobe is the pituitary. The pituitary receives blood from the main mass of the hypothalamus, and thus it is the first organ exposed to GRS (see Figure 8.3).

The pituitary is sensitive to GRS and reacts to its presence by producing and releasing two different hormonal substances known collectively as **gonadotropins.** These gonadotropins are secreted in the blood and are carried all over the body. Both sexes produce both kinds of gonadotropins, and both sexes react in similar ways to the presence of each. The first of the gonadotropins is called **follicle-stimulating hormone,** or FSH. FSH promotes meiosis or gametogenesis in both sexes. Its name is derived from the situation in females, where the follicle that contains an egg increases greatly in size. The second gonadotropin is called **luteinizing hormone,** or LH. LH affects the nonmeiotic cells in the gonads and causes them to produce their own array of hormones. In

Figure 8.3

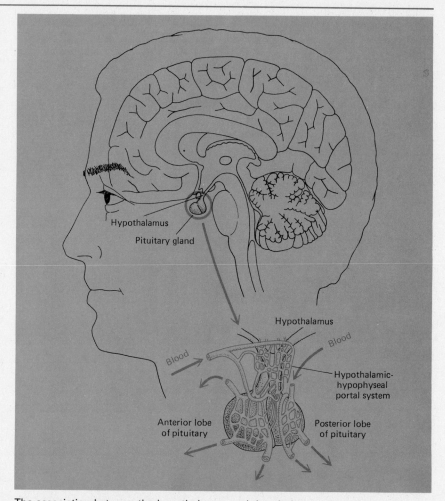

Hypothalamus

Pituitary gland

Hypothalamus

Blood

Blood

Hypothalamic-
hypophyseal
portal system

Anterior lobe
of pituitary

Posterior lobe
of pituitary

The association between the hypothalamus and the pituitary.

females LH stimulates the cells lining the follicle to produce female sex hormones, or **estrogens,** and also promotes the differentiation of some cells into a structure called the **corpus luteum** ("yellow body"). In males LH stimulates cells between the seminiferous tubules, the so-called **interstitial cells,** to produce male sex hormones, or **androgens.** Because the interstitial cells of males react to the presence of LH, this hormone is sometimes known as interstitial-cell-stimulating hormone, or ICSH. ICSH and LH are the same substance (see Table 8.1).

GONADS

Ovaries   There are several different estrogens produced by the ovaries, all of which tend to promote the development of those characteristics we think of as feminine. Small quantities of estrogen produced long before birth promote the differentiation and continued development of those organs directly concerned with the production and movement of gametes—the primary sexual characteristics. These include the ovaries and the associated tubes and glands. As girls approach 12 or 13 years of age, estrogen production increases under the influence of successive

Table 8.1
Table of Reproductive Hormones

ENDOCRINE GLAND	HORMONE	ORGAN(S) AFFECTED	EFFECT
Hypothalamus	GRS	Adenohypophysis	Produces gonadotropins
Anterior lobe of pituitary (adenohypophysis)	FSH	Ovaries and testes	Promotes meiosis
	LH (ICSH)	Ovaries and testes	Production of sex hormones
	LTH	Corpus luteum	Maintenance of uterine lining
		Mammaries	Lactation
Posterior lobe of pituitary	Oxytocin	Uterus	Increases muscular contractions
Ovary	Estrogens	Many	Differentiation and development of primary and secondary sexual characters
		Adenohypophysis	Inhibits production of FSH; produces LTH
Testes	Androgens (testosterone)	Many	Differentiation and development of primary and secondary sexual characters
		Adenohypophysis	Inhibits production of gonadotropins
Corpus luteum	Progesterone	Uterus	Maintains vascularization
		Adenohypophysis	Inhibits production of LH
Chorion and placenta	Chorionic gonadotropin	Ovaries	Production of estrogens
	Estrogens	Adenohypophysis	Inhibits production of FSH; promotes LTH secretion
		Pelvic girdle	Relaxation of pubic ligaments
	Progesterone	Uterus	Maintenance of pregnancy
		Adenohypophysis	Inhibits production of LH

increases in the levels of GRS and LH circulating in the blood. The increased estrogen levels promote the development of what are called secondary sexual characteristics. These characters are not directly concerned with the production or movement of gametes but seem to facilitate the recognition of the sexes as different.

One of the effects of estrogens is to accelerate growth in many different bones, which is the reason that preadolescent girls are frequently larger than boys of the same age. However, prolonged exposure to estrogens also promotes the rapid closure of the growth regions in the long bones. Thus girls generally stop growing earlier than do boys. The promotion of bone growth by estrogens is particularly apparent in the girdles of the shoulder and hip. Women generally have wider shoulders and hips, in proportion to height, than do men, which gives them a distinctly different body shape. This difference in skeletal structure is emphasized by the tendency of estrogens to stimulate the development of fatty tissue in regions such as the abdomen, hips, and buttocks. This not only gives women a softer, rounder appearance but also decreases their density and increases their capacity to retain body heat. Estrogens also suppress muscular development, but they promote the development of the breasts, or mammary glands (see Figure 8.4). Finally, high levels of estrogens suppress or inhibit the secretion of FSH, probably by affecting the hypothalamic output of GRS. This provides a **negative feedback** mechanism by which the hypothalamus, pituitary, and ovaries are prevented from overdoing a good thing.

Testes   The principal androgen produced by the testes is **testosterone.** Testosterone promotes the development of distinctive male features including both primary and secondary sexual characteristics. By approximately the seventh week after fertilization, small quantities of testosterone are promoting the differentiation and development of the male reproductive organs. By birth these organs are recognizably different from those of females; however, no further development appears to occur until adolescence. At that time the hypothalamus stimulates the pituitary, which causes the testes to start producing gametes and increase the output of testosterone. This upsurge in testosterone levels increases the size and the level of maturity of the male reproductive system.

Testosterone also accelerates the growth of bones, particularly the long bones of the limbs, causing adolescent males to undergo a characteristic growth spurt. As is the case for females, prolonged exposure to sex hormones promotes the closing off of the growth regions in these bones, thus terminating the growth spurt. However, the onset of the growth spurt is much later in males than it is in females. Thus both

Figure 8.4

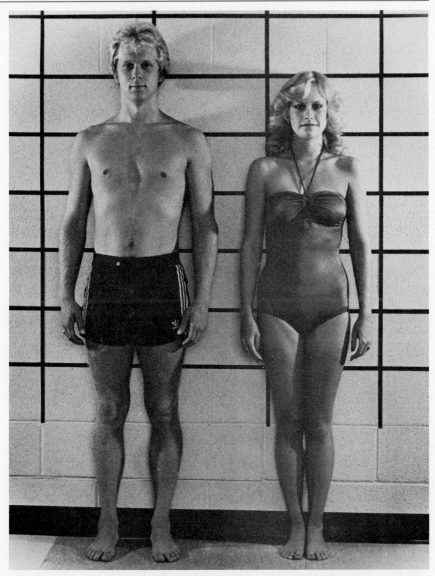

Demonstration of the differences in body proportions, muscular development, distribution of fat, mammary development, and so on, found between the sexes in people. [Photo courtesy of Matthew E. Stockard.]

sexes terminate their growth with one last surge, but the length of time devoted to growing is much longer for males. This generally means that males have proportionately longer arms and legs than females and thus are taller. Increased levels of testosterone circulating in the blood also promote an increase in the size of the voice box and vocal cords, the growth of body and facial hair, and the conversion of fat to protein in muscle tissue. This results in men having deeper voices, being hairier, and being physically stronger than women. Many of these differences are evident in Figure 8.4. There is also strong evidence from nonhuman vertebrates, and at least some strong implications from human studies, that testosterone has a tendency to promote aggressive behavior in males. Finally, high levels of testosterone have an effect on the hypothalamus, which results in a decrease in the production of pituitary gonadotropins.

HORMONES AND DIFFERENCES

It may be surprising to learn that the large differences between sexes with which we are familiar are brought about by such minor differences in internal chemistry. Most of the structural differences between men and women are caused by the differences in composition of the sex hormones and the times at which they are released. A comparison between molecules of an estrogen and androgen (see Figure 8.5) reveals that the differences between these hormones are subtle indeed. Another example of the subtlety of sexual distinctions is the already mentioned differences in the production and release of gametes.

It should be apparent that because of the negative feedback mechanism in males, a very delicate balance must exist in the production of

Figure 8.5

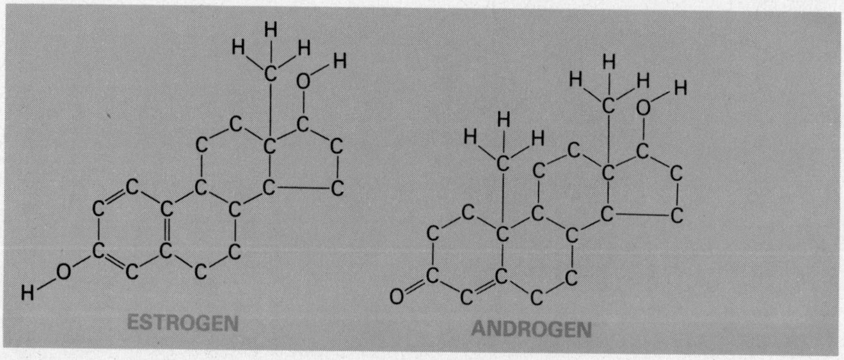

ESTROGEN   ANDROGEN

Comparison of an estrogen and an androgen (testosterone).

GRS, gonadotropins, and androgens. If the production of any of these hormones is too high, elevated levels of testosterone will result, which will cut back production in the whole system. Too low a production of any of these hormones will speed up the production of GRS, thus elevating the activity of the whole system. On the whole, the system shows a strong tendency to maintain a fairly constant level of all the relevant hormones. Because the hormone that regulates gamete production is FSH, a gonadotropin, and because the gonadotropin levels are maintained at a reasonably constant level by the negative feedback system, the production of gametes must be relatively constant.

Despite their possession of the same kind of negative feedback systems involving GRS, gonadotropins, and sex hormones, vertebrate females do not generally exhibit a constant production and release of gametes. The females of most vertebrate species are cyclic with respect to ovulation, the length of the cycles varying greatly among different species. Some species of rats and mice have cycles that last four or five days. Other species, such as deer, have cycle lengths of about a year. Presumably the negative feedback system in females works in a slightly different way than it does in males of the same species.

Human Menstrual Cycle

Most mammals—those warm-blooded vertebrate animals with hair—have what is known as an **estrous cycle. Estrous** is a word that describes what the female is like during that part of her productive cycle when ovulation occurs. One of the most noticeable symptoms of estrous is the presence of a vaginal discharge. During estrous the female behaves in a way that people usually interpret as indicating that she is very interested in reproductive activity—for example, we say that a cow, a dog, or a cat is "in heat" or "in season." Species with estrous cycles usually mate only when the female is in the estrous phase of her cycle. As mating then coincides with ovulation, it maximizes the chances of a successful fertilization. The duration of the estrous cycle is measured as the amount of time occurring between successive onsets of estrous, which is the same as saying the amount of time between successive ovulations.

Primates—that small group of mammals that includes monkeys, apes, and people—have a different kind of reproductive cycle, called a **menstrual cycle.** Menstrual cycles are also characterized by the periodic appearance of vaginal discharges, called **menstruation,** but this event does not coincide with ovulation. Repeated observations indicate that this periodic bleeding occurs approximately halfway between successive ovulations. By definition, the onset of menstruation constitutes the first day of the menstrual cycle. The regularity of cycle length among individual human females can vary widely, but, on the average, 28 or 29 days elapse between the onset of successive menstruations.

PREOVULATORY PREPARATION

On the first day of the menstrual cycle vaginal bleeding commences. This discharge is not really blood but a mixture of blood, mucus, and dead or dying **epithelial** cells from the inner wall of the uterus. Menstruation is actually the result of the periodic shedding of this inner surface of the uterus. The menstrual flow gradually tapers off and generally terminates by about the fifth day, although this too varies individually. During these initial days of the cycle, circulating levels of the gonadotropic hormones, FSH and LH, are very low. Apparently at this time the hypothalamus stimulates the anterior lobe of the pituitary to start producing the gonadotropic hormones. Over the next approximately ten days the concentration of these substances circulating in the blood increases substantially.

The increasing concentration of FSH causes one of the follicles in an ovary to start increasing in size. This multicellular structure contains the egg, or ovum, which will be the female gamete. The increasing concentration of LH causes the ovary to release larger and larger amounts of estrogens.

The estrogens have an immediate effect upon the wall of the uterus, causing them to thicken and acquire many small blood vessels. By day 10 or 12 of the cycle the uterine lining is a spongy mass, heavily infiltrated with thin-walled blood vessels. At approximately this point in the cycle the levels of circulating estrogens are high enough to turn off the production of FSH by the pituitary. The subsequent decline in the concentration of circulating FSH, together with a continued increase in the concentration of circulating LH, causes the swollen follicle to rupture and release its egg. This rupture of a follicle can sometimes be detected by the woman as a sharp internal twinge and may be accompanied by some vaginal spotting caused by small broken blood vessels in the ovary.

OVULATION

Ovulation is a critical time in the menstrual cycle because it is the only time when a viable female gamete is present. The ovum is quickly scooped up by the funnel-shaped upper part of the oviduct and moved steadily down the interior of the oviduct. Unless a sperm cell fertilizes the egg within approximately 48 hours, it dies and continues its trip all the way through the oviduct, uterus, and vagina.

At the time of ovulation several hormonal events occur almost simultaneously. The site of the ruptured follicle differentiates into the corpus luteum, which releases another hormone, called **progesterone.** Its chief effect is to maintain the spongy condition of the uterine wall. A number of follicular cells are destroyed during ovulation, which causes a surge of estrogens to be released. This surge of circulating estrogens has its most significant effect on the anterior lobe of the pituitary, which now releases a third hormone. The new pituitary hormone has been variously called **prolactin,** lactogenic hormone, mameogenic hormone, or **luteotropic hormone** (LTH) because of its different effects. Its most immediate effect is to maintain the status of the corpus luteum, which continues to produce progesterone. As the levels of circulating progesterone increase, they too affect the pituitary, causing it to cease the secretion of LH. As LH concentrations decline, the production of estrogens terminates.

NONFERTILIZATION AFTERMATH

If the egg released at ovulation is not fertilized within about two days, it dies. By the seventh day after ovulation—which is approximately day 21 in the menstrual cycle—the circulating levels of estrogens, FSH, and LH are all declining, or are at their minimum levels. Circulating levels of LTH and progesterone are still very high, but the decline in estrogens causes the pituitary to stop secreting LTH. A decline in the concentration of circulating LTH causes a deterioration of the corpus luteum, which stops the production of progesterone. Within a week the hormonal support for the maintenance of the uterine wall has declined to the point where the uterine wall starts to slough off, which signals the start of another round of menstruation (see Figure 8.6). With the lowering of both estrogen and progesterone levels, the hypothalamus produces more GRS, and thus the pituitary is stimulated to secrete gonadotropins—and another cycle starts.

FERTILIZATION AFTERMATH

Normally if fertilization does occur, the sperm meets the egg within the oviduct and the zygote is formed there (see Figure 8.7). This new diploid organism continues to be moved down the oviduct. As the fertilized egg is propelled along, it undergoes several mitotic replications, reaching the 32-cell stage within approximately three days. As mitosis continues, cellular replication tends to get out of phase, and the developing embryo starts to resemble a hollow sphere of cells within which an inner mass of cells will become the individual (see Figure 8.8, p. 234). This stage of development is called the **blastocyst.**

Upon reaching the uterus, the blastocyst burrows into the lining of the uterus, aided by the production of digestive enzymes. The lead-

Figure 8.6

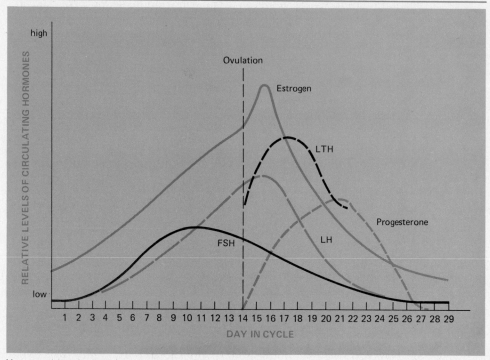

Hormonal levels during a normal menstrual cycle.

ing face of the blastocyst then grows a multitude of extensions that in-
sinuate themselves into the structure of the uterine lining. This step of
differentiation is important to the embryo for two reasons. First, it enables the new individual to make use of the mother's circulatory system—in a very real sense to act as a parasite. Second, these newly differentiated cells produce yet another hormone called **chorionic
gonadotropin**. Chorionic gonadotropin functions in much the same
way as LH, in that it promotes the secretion of estrogens by the ovaries.
The survival of the embryo is dependent on the release of chorionic
gonadotropin within several days after ovulation. The embryo's production of chorionic gonadotropin offsets the declining production of
LH by the maternal pituitary. This means that the circulating levels of
estrogens remain reasonably high, which maintains the pituitary's output of LTH. The LTH keeps the corpus luteum intact, which continues
to pump out progesterone, thus maintaining the integrity of the uterine
lining. Without the production of chorionic gonadotropin by approxi-

mately day 21 in the menstrual cycle, the various hormones associated with the maintenance of the uterine lining would decline to the point that menstruation would start on about day 28, and the embryo would be removed with the uterine lining.

The production of chorionic gonadotropin by the fingerlike projections into the uterus from the embryo continues for up to 20 weeks after fertilization. During this period the embryo undergoes a considerable amount of differentiation, including the development of the **placenta,** which is an elaborate connection to the uterine wall. By 20 weeks after fertilization the placenta starts producing large volumes of

Figure 8.7

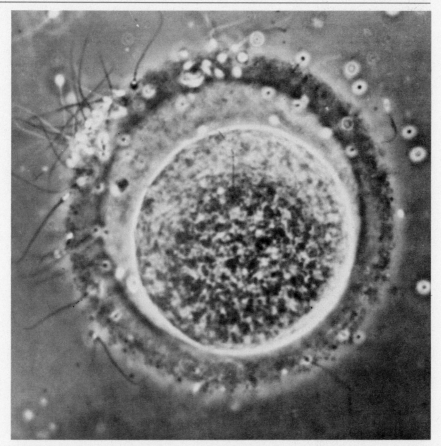

An egg and a sperm accomplishing fertilization. [Photo courtesy of Dr. Landrum B. Shettles.]

Figure 8.8

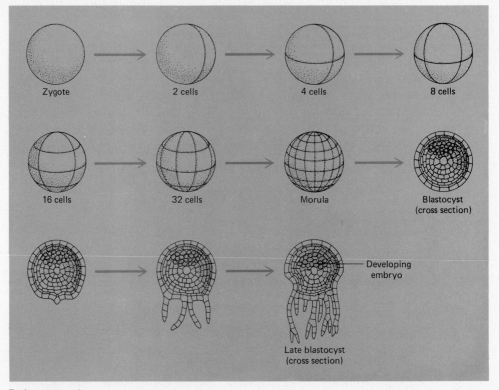

Early stages of embryonic development.

both estrogens and progesterone, which means that by this time the embryo itself is maintaining the uterus in the pregnant condition. Up until this point the mother has been supplying the estrogens and progesterone needed to maintain the uterus.

It is worth noting in passing that the blood of the developing embryo does not come in direct contact with that of its mother. Instead, the placenta consists of thousands of tiny blood vessels intermeshing over a large area on the inner surface of the uterus. While the cellular components of the two bloods remain separate, most other materials within the bloods are able to diffuse from one system to the other. Thus dissolved food molecules, water, oxygen, and vitamins can move from the mother to the developing embryo, and excess carbon dioxide, hormones, and nitrogenous wastes can move from the embryo to the mother's circulatory system.

While this system works fine for all parties concerned, the same principles of diffusion apply to any other dissolved materials present in either bloodstream. Any soluble materials will tend to move from the system where they are in higher concentration to the system where they are in lower concentration. Thus generally there is more of a possibility for movement of materials from the maternal bloodstream to the embryo's, for the simple reason that the mother has direct access to more chemical substances. Alcohol, nicotine, caffeine, and most drugs and medications can also diffuse across the placenta and produce effects on the embryo as well as on the mother. Some of these substances are known to interfere with normal developmental processes. Such interference can produce deleterious consequences of varying degrees of severity, from slight anatomical abnormality to embryonic death. For example, narcotic drugs taken by the mother can induce addiction and withdrawal symptoms in newborn infants, and thalidomide, a onetime sleeping pill, is known to interfere with the normal development of arms and legs by embryos.

As more and more time passes, the production of estrogens tends to exceed the production of progesterone to a greater and greater degree. By late pregnancy the level of circulating estrogens may be ten times higher than normal levels during the middle of a nonpregnancy menstrual cycle. The proportionately higher concentration of estrogens relaxes the **ligaments** binding the **pubic bones** in the hip girdle and promotes muscular contractions of the uterus. Another factor that stimulates the uterus to contract is the physical presence of a large object within the uterus. The longer the developing embryo remains within the uterus, the larger it becomes. A third factor involved in the normal termination of pregnancy is the posterior lobe of the pituitary. Using a mechanism that is poorly understood, the posterior lobe of the pituitary reacts to the fully developed embryo by releasing a hormone called **oxytocin.** Oxytocin, in conjunction with high estrogen levels, stimulates the uterine contractions of labor. The weakened ligaments of the pubic bones allow the pelvic girdle to separate enough for the infant to pass through the last portion of the female reproductive tract and into the world.

The birth of the infant and the placenta results in a sudden and precipitous drop in the circulating levels of estrogens and progesterone. LTH, which has been produced throughout the pregnancy, is now the only reproductive hormone found at any significant concentration. With the inhibitory effects of estrogens and progesterone removed, and with the presence of oxytocin as a supplement, LTH now serves as a stimulus for the ejection of milk by the breasts. The mechanical stimulation of suckling by the newborn sends a nervous impulse to the brain (hypothalamus), which continues to stimulate both lobes of the pituitary to release luteotropic hormone and oxytocin. These two hormones maintain the milk-producing capability of the mammary glands.

It appears that, at least to some extent, the reinforcement of nursing, LTH production, and **lactation** have an inhibitory effect on the hypothalamus's and pituitary's abilities to reestablish the normal cyclic production of FSH and LH. In many cases when a woman ceases nursing, there is a decline in the production of LTH, and shortly thereafter the normal menstrual cycle is resumed. There are also many cases in which pregnancy has recurred shortly after birth. This seems to be another example of the large amount of variability that can exist among individuals of the same species.

Living with the System

It may perhaps be belaboring the obvious, but with the information given, it should be apparent that the only practical way to achieve fertilization and pregnancy is to transport live sperm to a live egg. The reproductive systems of males and females have evolved together with all the secondary sexual characteristics to maximize the chances of accomplishing this unification. One can imagine all sorts of ways of bringing sperm and egg together, but the simplest and most effective way is to align the two ends of opposite tracts in such a way as to allow a continuous passage for the movement of sperm from the male system into the female system. In simple terms this process is called sexual intercourse. Then if an egg is encountered in the female system, fertilization and pregnancy is a highly probable outcome.

The various other methods proposed for accomplishing fertilization and pregnancy, in and of themselves, are highly ineffective. These include hand holding, kissing (of any ethnic persuasion), necking, petting, wearing knee socks, bikinis, or short skirts, smoking, drinking, loud singing (carousing), listening to raucous music, or reading any kind of written material. By the same token, many of the traditional folk methods for avoiding pregnancy, such as wearing a piece of tape over your navel, wearing lucky charms or amulets, having a tipped uterus, or praying devoutly, are all unlikely to be effective when practiced before, during, or after sexual intercourse. They can be effective if used instead of sexual intercourse.

The preceding paragraph has been written somewhat tongue in cheek, but the truth is that a lot of nonsense gets passed around on the subject of sexual reproduction. Many of the things mentioned may facilitate the development of the intimacy that leads to intercourse, some of them may enhance or call attention to sexual differences, but none of them result directly in fertilization and pregnancy.

If pregnancy is the natural result of the fusion of egg and sperm nuclei, which can only occur if live eggs and live sperm occur at the same place at the same time, then the chances of achieving pregnancy are maximized by having sexual intercourse at the time of ovulation. Conversely, the most effective means of minimizing the likelihood of pregnancy is to abstain from sexual intercourse at any time. Many people in our society favor this method—usually for people other than themselves.

CONTRACEPTION

One of the ways in which humans differ from many, if not most, other species is in their desire to engage in straightforward reproductive activity without any particular desire or intention of reproducing. There is good evidence to support the idea that engaging in this kind of behavior fulfills a social function whose ultimate purpose may be reproductive but not necessarily associated with any one particular act of sexual intercourse. While the rightness or morality of this human tendency may be subjected to extensive debate, the fact remains that a large number of people do it and are unpleasantly surprised when pregnancy becomes a real personal state of existence. As the number of unwanted or problem pregnancies continues to grow in this country, the associated social problems grow also. As a biologist, it seems to me that some of these problems could be avoided by a clear understanding of the normal functioning of human reproduction and some of the means that have been devised (by humans) to avoid pregnancy.

Periodic Continence   Undoubtedly the most effective means of avoiding pregnancy is to avoid sexual intercourse. In addition to being effective, this method entails no risk of moral censure from any organized social institution that I am aware of, and it severely limits the spread of **venereal disease** as well. Almost as effective as total abstinence in theory, and about as acceptable morally, is the idea of periodic continence—abstaining from sexual intercourse during the so-called fertile period. If fertilization can only occur if a live sperm and a live egg are together in the oviduct at the same time, and a live egg is only present during the 48 hours following ovulation, then fertilization (and pregnancy) should not be possible if there are no live sperm in the oviduct during that 48-hour period. Obviously this means no intercourse within 48 hours after ovulation, but it also means no intercourse for some period before ovulation to avoid the possibility of live sperm surviving to the time of ovulation. The life expectancy of sperm cells within the female reproductive tract appears to be less than 72 hours, and possibly as short as 36 hours. Prudence would seem to dictate the avoidance of intercourse for the three days preceding ovulation.

A knowledge of the timing of ovulation is critical in the effective practice of periodic continence. If a normal human menstrual cycle averages 28 days in duration, and ovulation precedes the onset of menstruation by about 14 days, then ovulation can be expected to occur, on the average, around day 14 of the cycle. If everyone worked like a statistical average, this would mean that avoiding intercourse during days 11 through 15 of the cycle would succeed in eliminating unwanted pregnancies all the time. However, as we have seen repeatedly, the world is full of all kinds of diversity, and not only do different women have cycles of different lengths, but the same woman can have cycles of different lengths at different times. To accommodate this variability in cycle length, we must expand the time period during which intercourse is avoided to extend from three days before the earliest possible time of ovulation to the second day after the latest possible time of ovulation. For some women with extremely variable cycles, this can effectively eliminate any possibility of intercourse without a high risk of pregnancy.

One means of eliminating some of the guesswork in pinpointing the time of ovulation is by keeping track of daily **basal-temperature** fluctuations. Contrary to popular belief, the human body does not maintain an absolutely constant temperature of 37°C (98.6°F). A person's basal temperature—that obtained from a person who has been resting quietly and not eaten for several hours, as, for example, before getting out of bed in the morning—can be seen to vary slightly from day to day. In women this variability in basal temperature has a pattern that correlates with the menstrual cycle, as can be seen in Figure 8.9. The preovulatory temperatures seem to fluctuate around an average. Immediately before ovulation the basal temperature makes a drop, followed by a sharp rise, and then fluctuates around a new, higher average basal temperature. At the onset of menstruation the basal temperature drops down to the preovulatory level. A record of basal temperatures for a normal menstrual cycle describes a pattern similar to a square root sign.

While this technique can be very useful for identifying when ovulation has occurred and, if menstrual cycles are very regular, for predicting when ovulation is most likely to occur next, it does not tell a woman with an irregular cycle that she is likely to ovulate three days hence. For women with irregular cycles the use of this technique may have its greatest value in identifying the time of ovulation responsible for the present pregnancy.

Figure 8.9

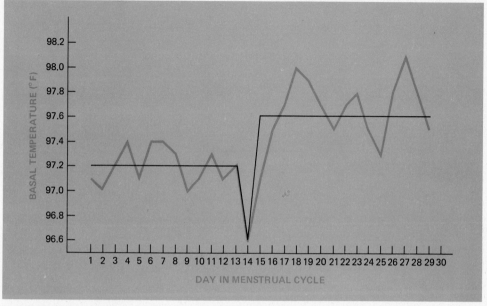

Daily basal temperatures.

There are several other difficulties with the practice of periodic continence. Many women report a feeling of increased desire, possibly caused by an upsurge in estrogens, at the time of ovulation. In a loving relationship this may not be an appropriate time to bury these emotional feelings. There also seem to be some indications that at least some women are induced to ovulate by the act of sexual intercourse. Timing intercourse to avoid ovulation in this situation does not appear to be possible. (One is led to wonder if the moral acceptance of periodic continence is in any way connected with its lack of practical effectiveness).

Physical Barriers The probability of achieving fertilization can also be reduced by interfering with the free passage of sperm through the female reproductive tract. This interference can be accomplished through either physical or chemical means. It is possible to use a **diaphragm** to wall off the vagina from the uterus, thus preventing sperm from reaching the oviduct and effecting fertilization. Similarly, the penis can be capped with a **condom,** or **prophylactic** device, which prevents the sperm from coming into contact with any part of the female reproductive tract. Chemical means of reducing sperm passage include the use of different foams, gels, and lubricants having spermicidal properties. When these are introduced into the vagina prior to intercourse, they provide a lethal environment for the sperm and thus reduce the chances of sperm successfully reaching the oviduct and accomplishing fertilization. The combined use of such physical and chemical barriers can be highly effective in avoiding pregnancy. However, all such barriers have been known to fail, and, even more frequently, people have failed to erect the barriers properly.

Hormonal Tinkering   The elegant system of hormonal control over the female menstrual cycle appears to be susceptible to modification by adjusting the levels of various hormones artificially. Ovulation, as we have seen, is precipitated by the increase in size of a follicle, promoted by FSH secretions, and high circulating levels of LH, which promote the rupture of that follicle. It has been noted for a long time that women who are pregnant do not ovulate. They do not ovulate because during pregnancy the woman's circulatory system is carrying high concentrations of both estrogens and progesterone, which have a negative effect upon the pituitary's production of both gonadotropins. Without these pituitary hormones the follicle does not grow and release an egg. This is the basic idea behind the functioning of the birth control pill, of which there are now probably several hundred varieties. Fundamentally, the pill is designed to put enough estrogens and progesterone into a woman's bloodstream to mimic the conditions of pregnancy and thus prevent the production of the pituitary gonadotropins and the resulting ovulation. If ovulation never occurs, it is exceedingly difficult to accomplish fertilization and real pregnancy.

While this simple approach is very effective for avoiding pregnancy, it does have some drawbacks. The pill does a very good job of simulating pregnancy, to the extent that many women experience many of the side effects of pregnancy, including morning sickness, water retention and bloat, breast swelling and tenderness, and weight gain. There is also some evidence that the spontaneous clotting of blood within the blood vessels occurs more frequently in women who are taking birth control pills. Such pills can only be taken with a doctor's prescription, and their effects should be monitored carefully by a competent physician. The long-term effects of such hormonal tinkering are still largely unknown, as these formulations have only been widely available since 1960.

Other Means   One of the most effective and widely used contraceptive methods, and probably the most poorly understood, is the intrauterine device, or IUD. The IUD consists of a small inert object, usually made of a soft plastic, which is placed inside the uterus. According to legend this method of contraception was first discovered by Arabic traders during the Middle Ages. Their chief method of transportation was by camel, and it was common knowledge that pregnant camels were less effective beasts of burden on a long journey than were nonpregnant camels. These traders discovered that if a few apricot pits were inserted into the camel's uterus, she would not become pregnant during the journey. Why this should be true is not clearly understood. Perhaps the physical presence of an object in the uterus stimulates more uterine contractions than is normal, and either this impedes the ability of sperm to reach the egg and accomplish fertilization or it propels the zygote through the uterus too quickly for it to implant. Whatever the explanation is, the use of the IUD is highly effective in preventing pregnancies.

As with all the other methods of contraception discussed, there are also drawbacks to the use of the IUD. Women carrying IUDs tend to have prolonged menstrual periods accompanied by heavier-than-normal bleeding and severe cramping. There have been reports of IUDs becoming embedded within the uterus and even puncturing the uterine wall. Some women appear to be more susceptible to infections of the reproductive tract while carrying an IUD, and there have been some reports of cancer associated with the use of IUDs. Because an IUD must be positioned by a physician, it is best to consult with your doctor if you are planning to use this approach to contraception.

Sterilization   By far the most drastic means of contraception, in terms of permanence, is **sterilization.** For people who have achieved their desired family size, this may be an ideal course of action. Sterilization involves the removal of pieces of tubule used to transport either sperm or eggs from the gonads.

In males sterilization is a relatively simple surgical procedure called a **vasectomy,** which can be performed in a doctor's office. It involves cutting out small pieces and tying off the ends of the tubes that run from the testes through the scrotum. This prevents sperm from being transported out of the reproductive tract and accomplishing fertilization. Spontaneous regeneration of the cut ends is highly improbable, so fertilization is not possible once the lower ends of the tubes have been cleared of sperm.

In females the operation is called a **tubal ligation.** Small pieces of oviduct are removed and the ends tied off, thus preventing an egg from passing to the uterus or sperm from reaching an egg and fertilizing it. Because the oviducts are located within the abdominal cavity, tubal ligations generally require major surgery.

It is worth mentioning that in most sterilization techniques the gonads are left intact, which means that they continue to produce gametes and perform their endocrine functions. Thus people who have been sterilized continue to look and act like normal males and females. The only real difference is that their gametes are unable to leave their respective reproductive systems, so they die there and are quietly resorbed. It is generally conceded that the reversibility of sterilization is exceedingly limited.

SEX DETERMINATION

As was demonstrated in Chapter 5, sex determination in humans is accomplished at the time of fertilization by the composition of the sex chromosomes in the involved gametes. All normal eggs carry an X chromosome, whereas sperm may carry either an X or a Y. Those royal males who have disposed of their wives because of an alleged reluctance to bear male offspring have probably not studied biology very carefully. Such cases serve as excellent examples of the intense interest that can be generated regarding the gender of proposed offspring. A natural question to ask is if anything can be done within reasonable limits to influence the determination of sex.

The usual and safest answer to such a question is, "Probably not." The available evidence seems to indicate that unless one resorts to some very complex technology using artificial insemination, the chances are about equal for an egg to be fertilized by an X-bearing sperm or a Y-bearing sperm.

However, there are some observations that enable us to generate a testable hypothesis. X chromosomes are significantly larger than Y chromosomes, which means that they should weigh more. This would imply that X-bearing sperm are heavier than Y-bearing sperm. Smaller, lighter Y-bearing sperm should be able to swim faster, but they would use up their energy reserves more quickly and die earlier than the larger, heavier X-bearing sperm. This situation could easily have an effect on the probabilities of sex determination if this chain of reasoning is correct. One would expect that when intercourse occurs shortly after ovulation, Y-bearing sperm are most likely to encounter an egg first because they move most rapidly. Consequently, the zygote formed at fertilization should have a chromosomal makeup of XY, which is male. Alternatively, when intercourse occurs well before ovulation, the long-lived X bearing sperm are most likely to be alive when the egg is released, and fertilization should result in a zygote with two X chromosomes, which is female. If most people practice intercourse without reference to the time of ovulation, it seems reasonable to assume that half the time it will occur before ovulation and the remaining half the time it will occur afterward, resulting in a 1:1 ratio of male to female offspring.

There is, to the best of my knowledge, no firm support for the validity of this particular hypothesis. By the same token, I am aware of no data that disprove its validity. However, it may not be practical to obtain the data necessary to rigorously test this particular generalization.

Summary

1 / *Human reproduction, like most living functions, is controlled and coordinated through an intricate system of biological regulatory mechanisms.*

2 / *The organs responsible for the production and movement of gametes in males and females are complementary to one another.*

3 / *While gamete production by human males tends to be constant, that of human females tends to be rhythmic or periodic.*

4 / *Gamete production is regulated by hormones produced in the endocrine glands.*

5 / *The three most important endocrine glands regulating reproduction in human beings are the hypothalamus, the pituitary, and the gonads.*

6 / *The sexual differences that we are aware of are generated by minor differences in the chemical composition of hormones and the time of release of those hormones.*

7 / *In animals with estrous cycles, vaginal discharges and sexual intercourse (copulation) coincide with ovulation.*

8 / *In animals with menstrual cycles, vaginal discharges occur approximately halfway between successive ovulations. Sexual intercourse may be independent of either of these two events.*

9 / *Increasing concentrations of circulating gonadotropins (FSH and LH) ripen the follicle for ovulation and cause the gonads to release increasing amounts of estrogens. Estrogens prepare the uterus for conception.*

10 / *If the released egg is not fertilized, the egg dies and passes through the female reproductive tract, all hormonal concentrations decline, and the uterine lining breaks down, thus starting another menstrual cycle.*

11 / *If the released egg is fertilized, it undergoes repeated mitotic replications and implants in the uterine lining. It also produces chorionic gonadotropin, which maintains high circulating levels of estrogens, LTH, and progesterone, and thus maintains the spongy condition of the uterine lining.*

12 / *The hormones of late pregnancy stimulate strong uterine contractions and a weakening of the connection of bones in the hip girdle, resulting in birth.*

13 / *Fertilization requires the presence of live egg and live sperm together in the same place (usually the upper reaches of the oviduct) at the same time. Properly timed intercourse maximizes the chances of successful fertilization.*

14 / *Techniques for minimizing the chances of fertilization include abstention, periodic continence, use of physical and/or chemical barriers, altering the normal hormonal changes with the pill, using an intrauterine device, or sterilization.*

15 / *A simple predetermination of sex in offspring is not yet possible.*

Questions to Think About

1 / What major biological activities do human beings have to perform in addition to reproduction?

2 / Why does the maturation of the human reproductive system occur so late in the development of the individual? Why does it take so long, in absolute terms, for a person to be biologically able to reproduce?

3 / How could you test the assertion that androgens tend to promote aggressive behavior? How could you control for culturally induced differences in aggressive behavior? What is aggressive behavior?

4 / Why do people have the particular kinds of secondary sexual characteristics that they do? Can you see any historical patterns that indicate that these characteristics are becoming more or less different?

5 / Can you think of any reasons why the hypothalamus should be ultimately responsible for regulating reproduction?

6 / Why does it seem to be easier to get pregnant than to avoid that condition?

7 / Is it possible to enhance the possibility of achieving pregnancy? If so, how?

Energetics
of Life

Introduction

In an earlier consideration of the distinctions between living and non-living, we noted that living things are composed of the same elements as nonliving things but that the proportions of those elements are quite different. This result implies that the arrangements of the elements as compounds are quite different. In fact, living things are usually composed of four classes of carbon-containing compounds known as carbohydrates, lipids, proteins, and nucleic acids. These compounds are uniquely associated with living material.

In the chapters on reproduction and heredity very strong evidence was offered to indicate that the hereditary material, the stuff of which genes are made, is the nucleic acid molecule DNA. If all living things can potentially reproduce, and by so doing pass traits onto their offspring, it seems most probable that all living things must have a supply of nucleic acids. This result has been found for every example of cellular life that has been investigated. From what has been shown about the way genes work, it would also be reasonable to conclude that all cellular organisms must possess a large number of proteins, both as enzymes and as structural components.

Both proteins and nucleic acids are large polymeric molecules with a great deal of structural specificity. The specific sequence of their respective monomeric units confers some information content to each of these kinds of molecules, which also implies a high level of organizational complexity.

I have alluded earlier to the idea that organizational complexity implies an energy input. Something that is complex and highly organized can conceivably occur on the basis of pure chance, but it is very improbable. It is even less likely that duplications of this same complex, highly organized entity could be brought into existence through the random interaction of chance events. This book is an example of a complex, highly organized entity in that it consists of a relatively small number of abstract symbols arranged in a specific sequence in which we perceive order. This order has been achieved through the expenditure of energy. There is a remote chance that an infinite shuffling of letters and spaces could result in the production of the same order, but the probability of achieving hundreds of copies of this same order by purely chance events is not worth considering. The same thing must be true for the vastly more complex order that we can perceive in the organization of a living cell. For the production of living organisms it is necessary to have the basic raw materials and a source of energy that can put those raw materials together in an appropriate way.

An extremely relevant question to ask at this point is: What is the source of the raw materials and energy used in producing living materials? Because the nonliving material world contains all the elements used by living cells, it seems reasonable to assume that this is one source of the raw materials. We can also see that many forms of life consume other forms of life, implying that many living things find their raw materials from other parts of the living world. Human beings are an excellent example of this observation. The raw materials used to produce more and more human cells, resulting in both individual growth and the reproduction of new individuals, are supplied by those substances we ingest as food. You are probably also aware that the same food substances that are a source of raw materials in the construction of more human beings are also the sources of energy used to drive the various human functionings.

Generally speaking, organic compounds require energy to be constructed or synthesized. Implicit in that generalization is the idea that the energy involved in putting such a molecule together is trapped, that it is required to hold the molecule together. Conversely, if the molecule is somehow encouraged to fall apart or break down, the energy involved in its maintenance should be released. In theory, the energy released from such a molecular breakup can be used to synthesize some other molecule or to perform some other function. The most familiar examples of energy release in the breakup of organic molecules are the combustion of such materials as coal, petroleum, wood, or paper.

Reconsidered in this new light, food becomes any organic compound that an organism can break down into more fundamental units, thus releasing energy in some usable form. Even a momentary reflection on that statement is enough to suggest the diversity of food substances that are available to different living organisms. However, the specification of organic compounds requires that only living things produce the food molecules that are needed to produce more living material. Ultimately, any conscientious observer will notice that the overwhelming majority of food substances are being produced by green plants and various kinds of greenish protistans and monerans.

Photosynthesis

Apparently there is something special about green plants and plantlike organisms. They do not take in food the way we expect organisms to do, and yet they continue to increase in size and numbers in characteristic living fashion. This simple observation has been made repeatedly for thousands of years. In spite of the antiquity of this observation, the earliest indications of serious study of the subject date back only to the

Figure 9.1

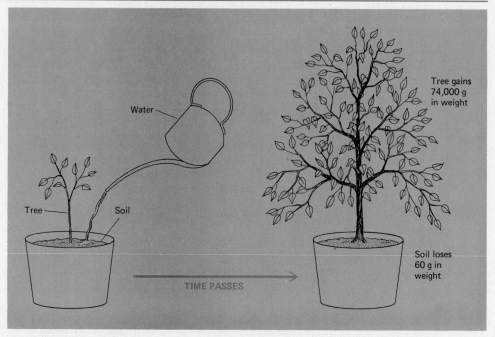

Water

Tree   Soil

TIME PASSES

Tree gains
74,000 g
in weight

Soil loses
60 g in
weight

Van Helmont's experiment.

mid seventeenth century. At that time Jean Baptiste van Helmont reported on a set of observations on plant growth. He had planted a willow sprout of known weight in a tub of soil, whose weight was also known. For five years van Helmont carefully tended his willow by sprinkling it with either rainwater or distilled water, but he added nothing else and made certain that no soil was ever removed from the tub. Finally, he uprooted the plant and weighed it and the soil (see Figure 9.1). The plant had gained approximately 74 kilograms in weight while the soil had lost approximately 60 grams. It was obvious that the weight gained by the plant had not been derived exclusively from the soil. Not having controlled any other variables, van Helmont concluded that the gained weight of the plant must be due to the added water.

By 1772 Joseph Priestley was able to add some additional observations to the properties of plant life. Priestley knew that a mouse kept in a sealed jar would suffocate and that a lit candle in a sealed jar would go out. He reasoned that in both of these cases some ingredient of air necessary to sustain mice and burning candles was being removed. Because a burning candle placed in a container where a mouse had previously suffocated snuffed out almost immediately, Priestley concluded that the ingredient needed was the same for both candle and mice. He expressed great surprise, however, when he observed that a sprig of green plant did not die under these circumstances. In fact, he observed that a green plant introduced into a sealed jar whose air had been previously exhausted by a lit candle would not only survive but would restore the air so that it could again support the burning of a candle. Mice placed in sealed jars suffocated, but mice placed in sealed jars with green plants survived (see Figure 9.2).

Figure 9.2

A lit candle kept in a sealed jar goes out

A live mouse kept in a sealed jar dies

A green plant kept in a sealed jar does not die

A live mouse kept in a sealed jar with a green plant does not die (until the mouse eats the plant)

Priestley's experiments.

Within a few years Jan Ingen-Housz demonstrated that Priestley's observations were valid only in sunlight and only for the green parts of plants. Plants kept in darkness exhausted the air in the same way that a candle or a mouse did. Nongreen parts of plants, such as roots, woody stems, flowers, and fruits, used up the unknown substance, even in the sunlight, just as candles or mice do.

By the early nineteenth century it had been determined that the substance removed from the air by animals, plants, and burning candles is oxygen. The green parts of plants, in sunlight, somehow manage to add oxygen to the air and at the same time make themselves heavier. This process requires the removal of another substance from the air, which we now call carbon dioxide. Since much of the weight gain in plants occurs in the form of carbon-containing organic molecules, it was reasoned that the carbon dioxide must be used to make the new organic compounds. Thus the whole process can be summarized by the following equation:

$$\text{carbon dioxide + water} \xrightarrow[\text{light}]{\text{green plants}} \text{organic compounds + oxygen}$$

or

$$CO_2 + H_2O \xrightarrow[\text{light}]{\text{green plants}} (CH_2O) + O_2$$

In chemical equations of this kind the arrow can be read as "yields" or "produces." The symbol (CH_2O) is a generalized formula for carbohydrates, which form the bulk of the organic compounds found in plants. This symbol represents the relative proportions of the included elements but does not reveal the actual size of the carbohydrate. For example, the simple six-carbon sugar glucose is usually represented as $C_6H_{12}O_6$ and ribose is symbolized as $C_5H_{10}O_5$.

This process, which builds up, or synthesizes, organic compounds in the presence of light, is called **photosynthesis.** The easiest interpretation of what happens during photosynthesis is that light energy is used to split off the two atoms of oxygen from the carbon dioxide, and then the water is added to the carbon as the basic unit in an organic compound. This means that the oxygen released in photosynthesis must come from the carbon dioxide. Photosynthesis is an excellent example of Murphy's law: "Nothing is as easy as it looks."

By the early part of the twentieth century there were indirect suggestions that photosynthetic oxygen was not released from the carbon dioxide. With the use of radioactive **isotopes** of elements as chemical tracers, it became possible to prove that this hypothesis was false. In the experiment a photosynthesizing plant is exposed to an atmosphere in which the carbon dioxide contains radioactive oxygen. Some of this so-called labeled oxygen appears in the resulting organic compounds, and the rest shows up in water vapor. None of it appears as molecular oxygen gas. If, instead, a photosynthesizing plant is exposed to normal carbon dioxide and water containing the labeled oxygen, the radioactivity appears in the released molecular oxygen. Thus the summary equation of photosynthesis must be rewritten as follows:

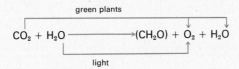

$$CO_2 + H_2O \xrightarrow[\text{light}]{\text{green plants}} (CH_2O) + O_2 + H_2O$$

It is necessary to place water on both sides of the equation, because the water produced at the end of the reaction is composed of different atoms than those in the water molecules at the beginning of the reaction.

This summary equation is sufficient to give you an overall idea of what occurs during photosynthesis. In actuality, the process is composed of at least two major parts: one concerned with trapping the energy of sunlight and the other involving the synthesis of the organic compounds. Both of these parts consist of a multitude of small-step chemical reactions that are frightfully complex. Since many of the details of each of these major parts of photosynthesis are still imperfectly understood, and since the detailed mechanisms by which the process is accomplished are not really necessary to gain an appreciation of the results of the process, I will forego an involved chemical description.

However, it is very important to understand all the information that is contained within the simple summary equation. It is important to stress the significance of the observation that only green plants—or, more specifically, the green parts of plants and other plant-like organisms—can carry out photosynthesis. The common feature possessed by photosynthetic organisms is a green pigment called **chlorophyll.** It is the pigment that enables its possessors to catch and store the energy in sunlight. Second, the process is dependent on sunlight, or, more specifically, those wavelengths of light energy present in sunlight that will excite the chlorophyll molecules. For the most part these wavelengths of light energy are not present in significant amounts in normal electric light bulbs or in moonlight. In addition, the photosynthetic process builds up large complex organic compounds, molecules that contain energy trapped in their chemical structures. Finally, photosynthesis is the only natural process that releases significant amounts of molecular oxygen as a waste product.

The photosynthesizers are (almost) unique in their ability to use a nonliving energy source to put water and carbon dioxide together into large organic compounds. The organic compounds manufactured by photosynthesis are *the* source of raw materials and energy needed to sustain (almost) all living organisms. The parenthetical qualifiers used in the preceding sentences are in recognition of the existence of some relatively insignificant forms of bacteria that sustain themselves without recourse to either photosynthesis or the products of photosynthesis.

Energy Extraction from Organic Compounds

It should be apparent that all living organisms must have one or more mechanisms by which they can extract for their use the energy trapped in organic compounds by photosynthesis. This has to be true for both the photosynthesizers and all those other organisms that either directly or indirectly consume photosynthesizers. The processes by which energy is extracted from organic compounds in a form that is usable by living organisms have been given several different names.

Under conditions where molecular oxygen is not present, as in deep-sea sediments and in certain soil habitats, energy can be derived from organic molecules through **glycolysis** and **fermentations.** As anaerobic conditions are presumed to have existed during the times when cellular life originated, glycolysis and fermentations are considered to be the oldest forms of energy extraction. Many organisms have achieved the additional capacity to extract more energy from a given organic molecule by coupling glycolysis to a process called cellular **respiration.** This latter process uses (and requires) molecular oxygen. Aerobic respiration is believed to be a more recent process that elaborates and refines the ancient process of glycolysis.

All these processes are **intracellular,** which means that they occur within cells. Unfortunately, the more familiar usage of the word *respiration* involves several other processes that are not intracellular and thus are quite different from the meaning of *cellular respiration* as used here. This conflict has caused more confusion in communication than is necessary. The following section is an attempt to clarify this potential source of confusion.

ELIMINATION OF RED HERRINGS

As you read this text in front of you, you may be vaguely aware of a rhythmic rise and fall of your chest accompanied by the passage of air through your nose and/or mouth. This process is called **breathing.** It is a mass movement of air in and out of some specialized organs called

lungs. Many different species engage in analogous processes whereby large volumes of gases, sometimes dissolved in water, are physically moved into the body. Fish pump water through their mouths and over their gills; some species of turtles pump water in and out of their bladders; insects draw air in and out of tiny tubules that ramify throughout their bodies. None of these things are cellular respiration. They can all be considered breathing.

The cells lining the lungs are very thin, and immediately behind them there are many very small blood vessels. Some gas molecules, such as oxygen, move out of the air in the lungs into the blood flowing through these small vessels. At the same time other gas molecules, such as carbon dioxide, move out of the blood into the air in the lungs. This phenomenon is called **gas exchange.** It is a passive, physical process caused by differences in the relative concentrations of the two gases in the air and in the blood. As long as the concentration of oxygen in the air of the lungs is greater than the concentration of oxygen in the blood, the oxygen will tend to move into the blood. Conversely, as long as the concentration of carbon dioxide in the blood is greater than the concentration of carbon dioxide in the air of the lungs, there will be a strong tendency for a net movement of carbon dioxide from the blood to the air of the lungs (see Figure 9.3). You will notice that there is no mention made of the body's need for oxygen or the necessity of getting rid of carbon dioxide. Gas exchange, whether it occurs in lungs, gills, or the tracheal tubes of insects, is a process governed by only the laws of **diffusion** and not by the needs of the organism.

It should be noted that both breathing, in the broadest sense of the term, and gas exchange are important adjuncts to cellular respiration in many organisms. They are means by which reactants and waste products of aerobic respiration are moved to and from the cells where aerobic respiration does occur. Because respiration occurs within cells, all organisms that respire must also carry out gas exchange. However, all organisms do not have to breathe. Many small aerobic organisms respire by using gases exchanged through their external body surfaces only.

ENERGY CURRENCY

Energy is the ability to do **work.** Work involves moving anything, from a mountain to a molecule. While energy is available in many forms, all of them are not equally useful in accomplishing a particular form of work. For example, a blowtorch releases a large amount of energy in the form of heat, which is not particularly useful for moving a large mass of material such as a rock or a brick. On the other hand, the mechanical energy of a driving piston or a rotating turbine is not the most

Figure 9.3

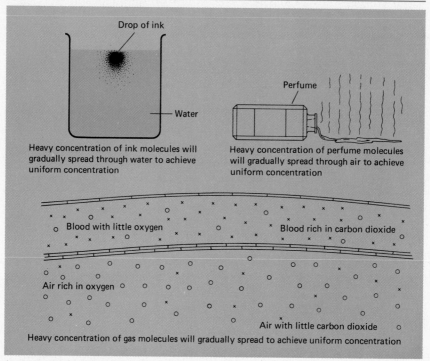

The principle of diffusion.

useful means of melting certain metals or converting water into steam. Most of the work done by living systems involves the transformation of chemical energy, and the form in which this occurs is a molecule called **adenosine triphosphate,** or ATP.

ATP is a compound of the nitrogen base adenine, the five-carbon sugar ribose, and three phosphate molecules arranged as shown in Figure 9.4. The connection between the second and third phosphates indicated by the wavy line is a relatively unstable high-energy **bond.** This means that when the third phosphate is removed from an ATP molecule, some energy is released, which can do molecular work. Of course, if an ATP loses its third phosphate, what remains is a molecule of phosphate and a molecule of adenosine diphosphate, or ADP. The conversion of ATP to ADP and phosphate is such a common means of powering the chemical reactions of living systems that it is known as the universal *energy currency* of living things. If molecular work is being done by a living organism, you can count on some ATPs losing a phosphate. Conversely, energy can be put into a usable form by generating a molecule of ATP. The easiest way to do this is to add a third molecule of phosphate to a molecule of ADP.

The use of the term *energy currency* as an allusion to the human use of money as a medium of exchange is an extremely useful device. If you, as a member of a reasonably complex human society, wish to get a particular piece of work done, the usual method is to spend some money so that the job will be done by someone skilled at the task. In the same way, a cell will spend some ATP to get some biological work accomplished. But how do you get the money in the first place so you can spend it to get work accomplished? For the most part you have to go out and do some work for someone else, who will then give you money for doing the work. For cells the ATP comes from the energy extracted from organic molecules used to put a third phosphate on a molecule of ADP.

The most important outcome of the following processes is that the energy trapped in organic molecules is used to generate molecules of ATP.

GLYCOLYSIS

The end product of photosynthesis is usually considered to be a simple carbohydrate such as the six-carbon sugar **glucose.** Glucose can serve

Figure 9.4

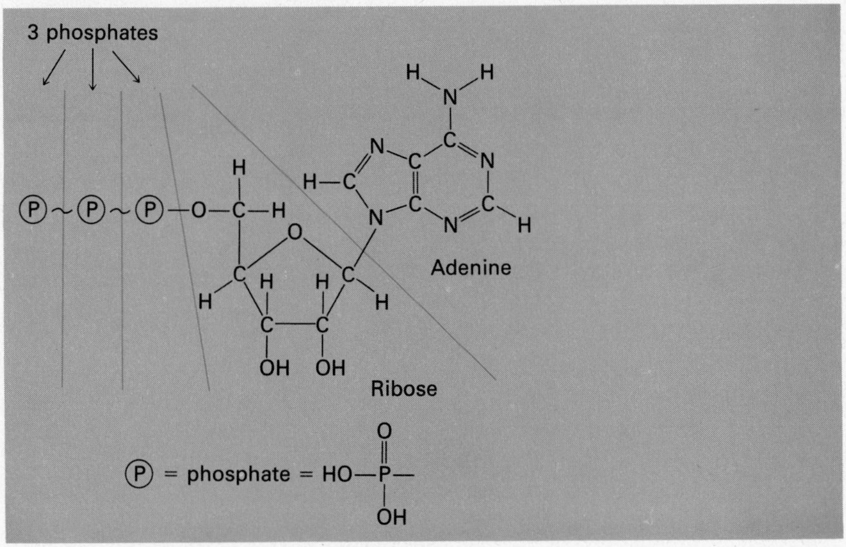

A molecule of adenosine triphosphate (ATP).

as a convenient starting point in a consideration of enery extraction. In extracting the energy tied up in a glucose molecule, all organisms start off by investing two molecules of ATP. The two phosphates from the two molecules of ATP are added to the glucose to produce a new, highly unstable, six-carbon molecule. This new molecule rapidly splits into two molecules, each composed of three atoms of carbon. These three-carbon molecules are called phosphoglyceraldehyde, or PGAL, each one of which contains one of the two phosphates from the two invested ATPs (see Figure 9.5).

Through a series of chemical reactions each molecule of PGAL is rearranged structurally into a molecule of pyruvate (see Figure 9.6). Each molecule of pyruvate also has three carbon atoms, so there has been no change in the number of carbon atoms involved. However, the structural rearrangements do cause the loss of several other atoms, resulting in the pyruvate's containing less total energy than was present in the PGAL. Each transformation of PGAL to pyruvate results in the generation of two molecules of ATP, and since one molecule of glucose produces two molecules of PGAL, then two transformations of PGAL to pyruvate will yield four molecules of ATP. However, it was necessary to use two molecules of ATP split the original glucose into two molecules of PGAL. This means that by the time the glucose molecule has been converted into two molecules of pyruvate, there are two more molecules of ATP than there were in the beginning. This can be put in more economic-sounding terms by saying that the conversion of one molecule of glucose to two molecules of pyruvate gives a net yield, or profit, of two molecules of ATP.

The conversion of a molecule of glucose to two molecules of pyruvate is called glycolysis. Glycolysis is the first step in any mechanism of converting energy to a biologically usable form. It is seldom the last step. What happens after glycolysis depends on several different factors, including whether or not there is molecular oxygen present and what kinds of enzymes the organism is capable of producing. One important thing to notice in the reactions described in Figures 9.5 and 9.6 is that molecular oxygen has not been involved. Because there has been no use made of molecular oxygen, glycolysis is described as an anaerobic process.

Another important outcome of glycolysis is the removal of some atoms of hydrogen. In the transformation of PGAL to pyruvate, two atoms of hydrogen are removed, meaning a total of four hydrogens are removed from the original glucose. Each pair of hydrogen atoms removed is picked up by a hydrogen-acceptor molecule called nicotinamide-adenine dinucleotide, or **NAD.** NAD is a derivative of one of the **vitamins** in the B_2 complex, **niacin** (see Figure 9.7). Once the NAD has

Figure 9.5

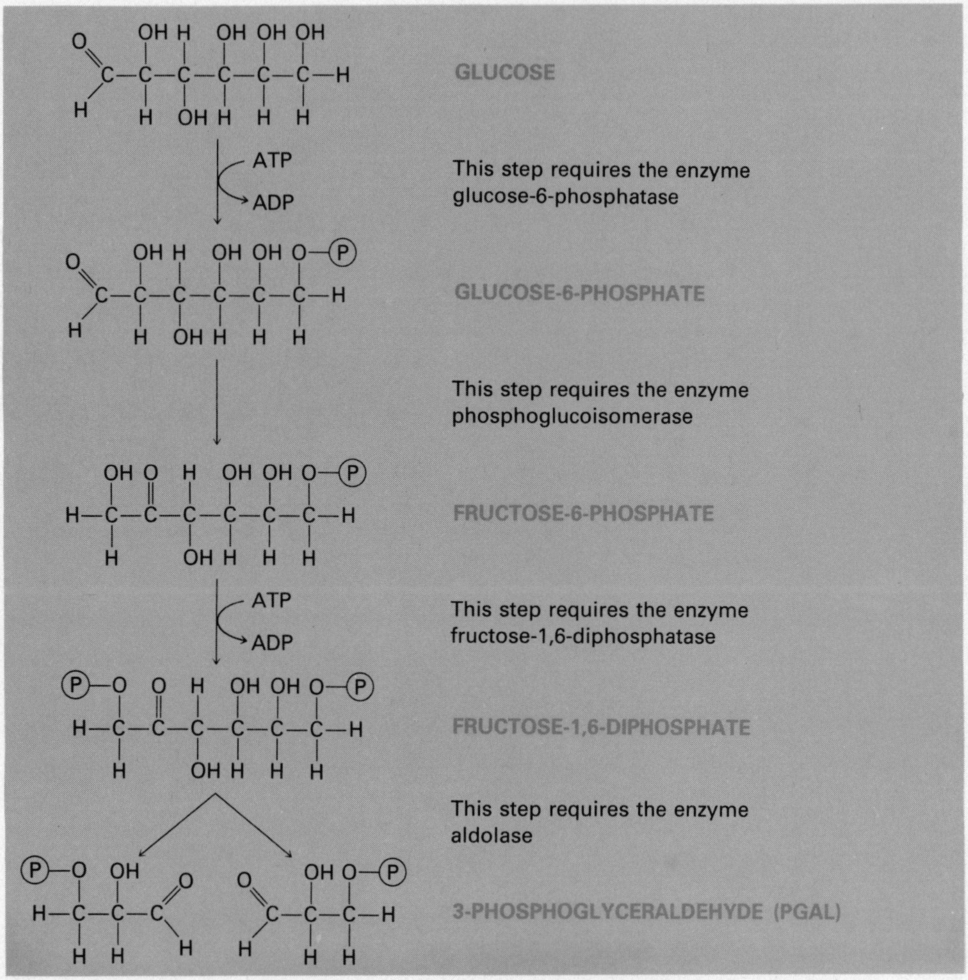

Conversion of a molecule of glucose to two molecules of PGAL.

picked up the two hydrogen atoms, it is known as reduced NAD, or NADH$_2$. Thus glycolysis yields two molecules of pyruvate, two molecules of reduced NAD, and a net gain of two molecules of ATP (see Figure 9.8).

FERMENTATION

As glycolysis continues, greater and greater quantities of both pyruvate and NADH$_2$ are produced. In the real world one seldom finds signifi-

Figure 9.6

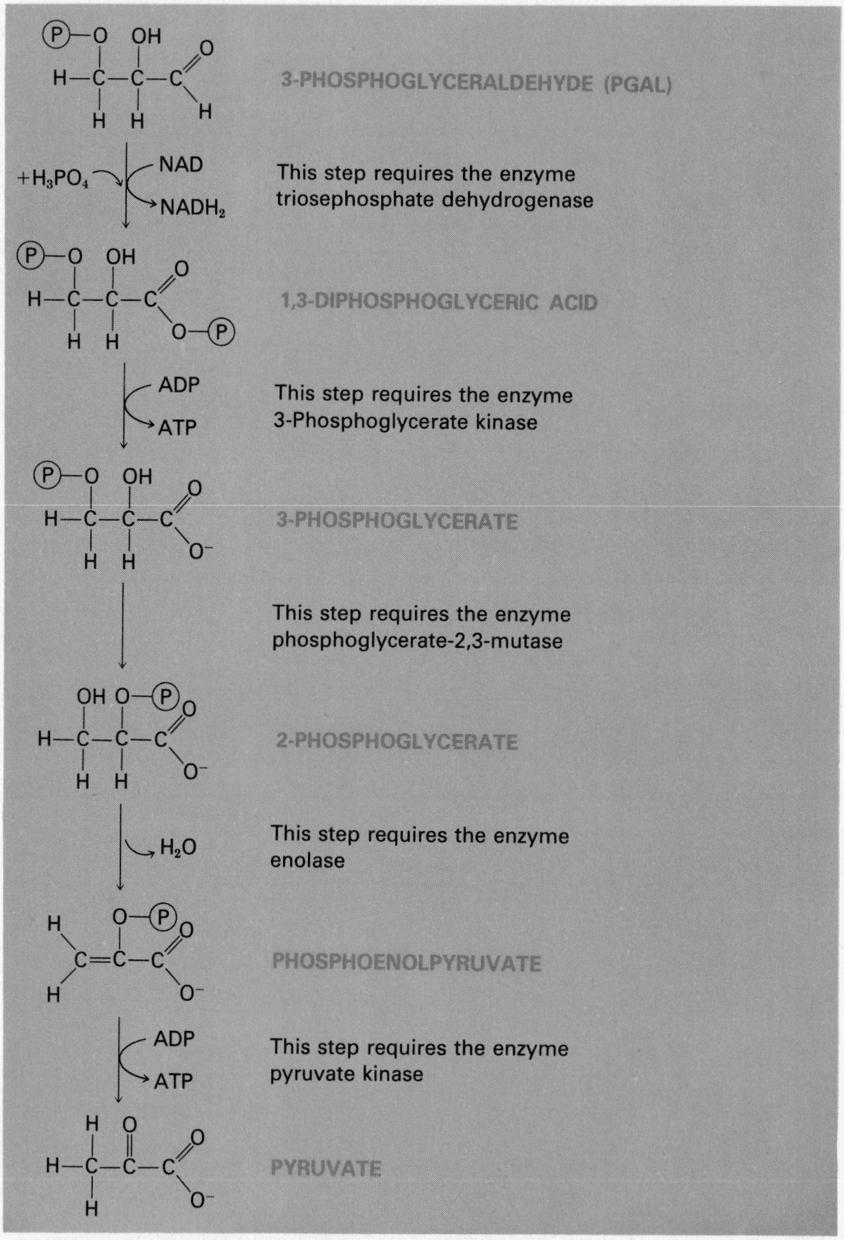

Conversion of a molecule of PGAL to a molecule of pyruvate.

Figure 9.7

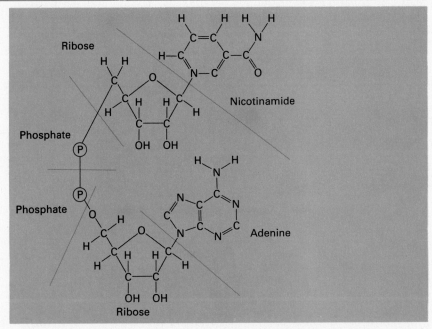

A molecule of nicotinamide-adenine dinucleotide (NAD). This molecule can be converted to $NADH_2$ by adding a pair of hydrogen atoms to the nicotinamide ring.

Figure 9.8

A simple schematic diagram of glycolysis.

cant amounts of either of these two products. If glycolysis has taken place under truly anaerobic conditions, the $NADH_2$ and the pyruvate tend to react with one another. The results of this reaction are very different, depending on the kinds of organisms in which the reaction occurs, which probably means that different kinds of organisms produce different kinds of enzymes. In animals and some kinds of microorganisms the $NADH_2$ gives up its hydrogens to the pyruvate, producing **lactate** and NAD. The NAD can, of course, cycle back and pick off more hydrogens to convert more PGAL to pyruvate. In plants and some other kinds of microorganisms the $NADH_2$ gives its hydrogens to the pyruvate, producing ethyl alcohol (also known as **ethanol**) and carbon dioxide as well as the expected molecule of NAD. No ATP molecules are produced in either of these conversions (see Figure 9.9). Both of these conversions of pyruvate have been called **fermentation.**

In summary, then, under anaerobic conditions some kinds of organisms can convert glucose to two molecules of lactate, for a net yield of two molecules of ATP. Other kinds of organisms, under the same anaerobic conditions, can convert a molecule of glucose into two mole-

Figure 9.9

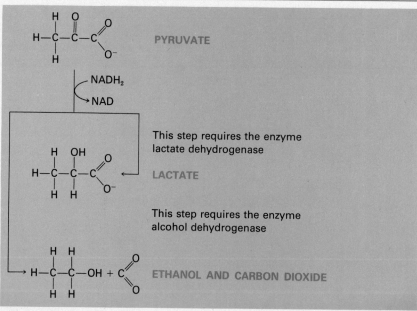

Fermentation.

Figure 9.10

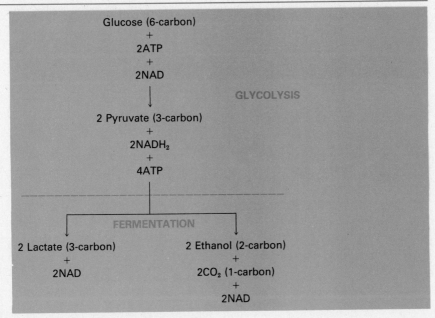

Summary of anaerobic energy extraction.

cules of ethanol and two molecules of carbon dioxide, for a net yield of
two molecules of ATP (see Figure 9.10).

AEROBIC RESPIRATION

If molecular oxygen is present, and if the cell possesses an appropriate
set of **electron transport** molecules, an entirely different outcome results. The hydrogens carried by the $NADH_2$ can be added to an atom of
oxygen to form water. This reaction releases a great deal of energy and,
as a consequence, is usually done very indirectly. Many organisms possess a series of molecules with a strong affinity for hydrogens, which
constitutes an electron transport chain. In eucaryotic cells the molecules associated with the electron transport chain are almost invariably
found within the mitochondria (see Figure 9.11).

   Under aerobic conditions an $NADH_2$ produced by glycolysis do-
nates its hydrogens to a molecule called flavin adenine dinucleotide, or
FAD. The hydrogens are then transferred to one **cytochrome** molecule
after another in a series, with a progressive decrease in energy. The energy released with each transfer of the hydrogens is used to add a phos-

Figure 9.11

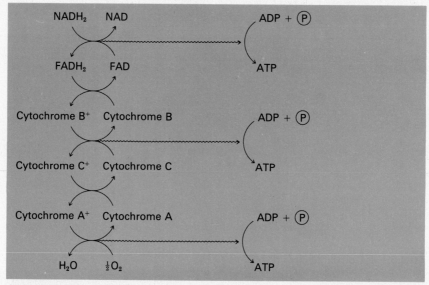

Electron transport chain. $NADH_2$ can eventually donate two of its hydrogens to oxygen, yielding a molecule of water. The energy derived from this process can be used to generate three molecules of ATP.

phate molecule to a molecule of ADP, thus producing ATP. The last cytochrome molecule in the series adds the hydrogen atoms to an oxygen, producing water. This whole process of moving the hydrogens from $NADH_2$, through the series of cytochromes, and adding them to water results in the production of three molecules of ATP. It also releases a molecule of NAD, which can return to the process of glycolysis, converting more PGAL to pyruvate.

Because glycolysis of one glucose molecule results in the production of two molecules of $NADH_2$, the presence of the cytochrome series and molecular oxygen allows for the generation of six more molecules of ATP. Together with the net gain of two molecules of ATP from glycolysis itself, there is now a net gain of eight molecules of ATP from one molecule of glucose.

With the hydrogens of $NADH_2$ being picked up by the cytochrome series, it is no longer possible for the $NADH_2$ to react with the pyruvate. Instead the pyruvate moves into the mitochondrion and splits into a molecule of carbon dioxide and a two-carbon molecule similar to acetic acid. This molecule attaches to a molecule of **coenzyme A,** forming something called acetyl CoA (see Figure 9.12). Coenzyme A is a deriva-

tive of another one of the vitamins in the B_2 complex, **pantothenic acid.**
The conversion of pyruvate to carbon dioxide and acetyl CoA releases
two more atoms of hydrogen, which are picked up by NAD, forming
$NADH_2$. If 2 molecules of pyruvate have been produced by the glycolysis of glucose, then 2 molecules of $NADH_2$ can be produced in converting those pyruvates into acetyl CoA. Each $NADH_2$ can be passed
through the cytochrome series to produce 3 more molecules of ATP, for
a net yield of 14 molecules of ATP from one molecule of glucose.

THE KREBS CYCLE

The final step of aerobic respiration involves the release of the energy
trapped within the two-carbon acetyl part of acetyl CoA. It is perhaps
one of the most astounding series of reactions ever discovered in that it
produces one of the original reactants with the release of a tremendous
amount of energy in usable form.

    The acetyl CoA reacts with a four-carbon molecule, called oxaloa-
cetic acid, to produce a six-carbon molecule, called citric acid, and to
release the coenzyme A molecule. In effect what happens is that the
coenzyme A transports the two-carbon acetyl to the four-carbon molecule and attaches them to produce a six-carbon molecule. Through an

Figure 9.12

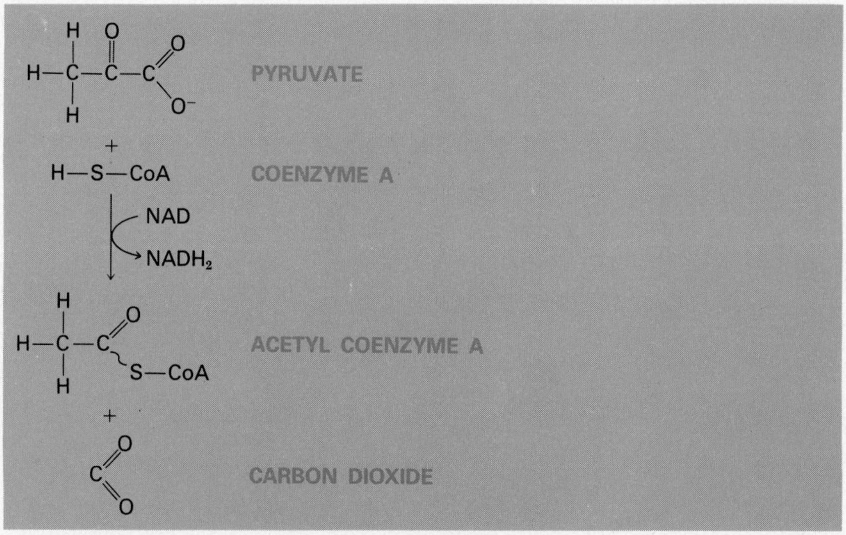

Conversion of pyruvate to carbon dioxide and acetyl CoA. The complex molecule of
coenzyme A has a reactive group consisting of one hydrogen and one sulfur (written as
H and S, respectively). The remainder of the molecule is represented best as CoA.

Figure 9.13

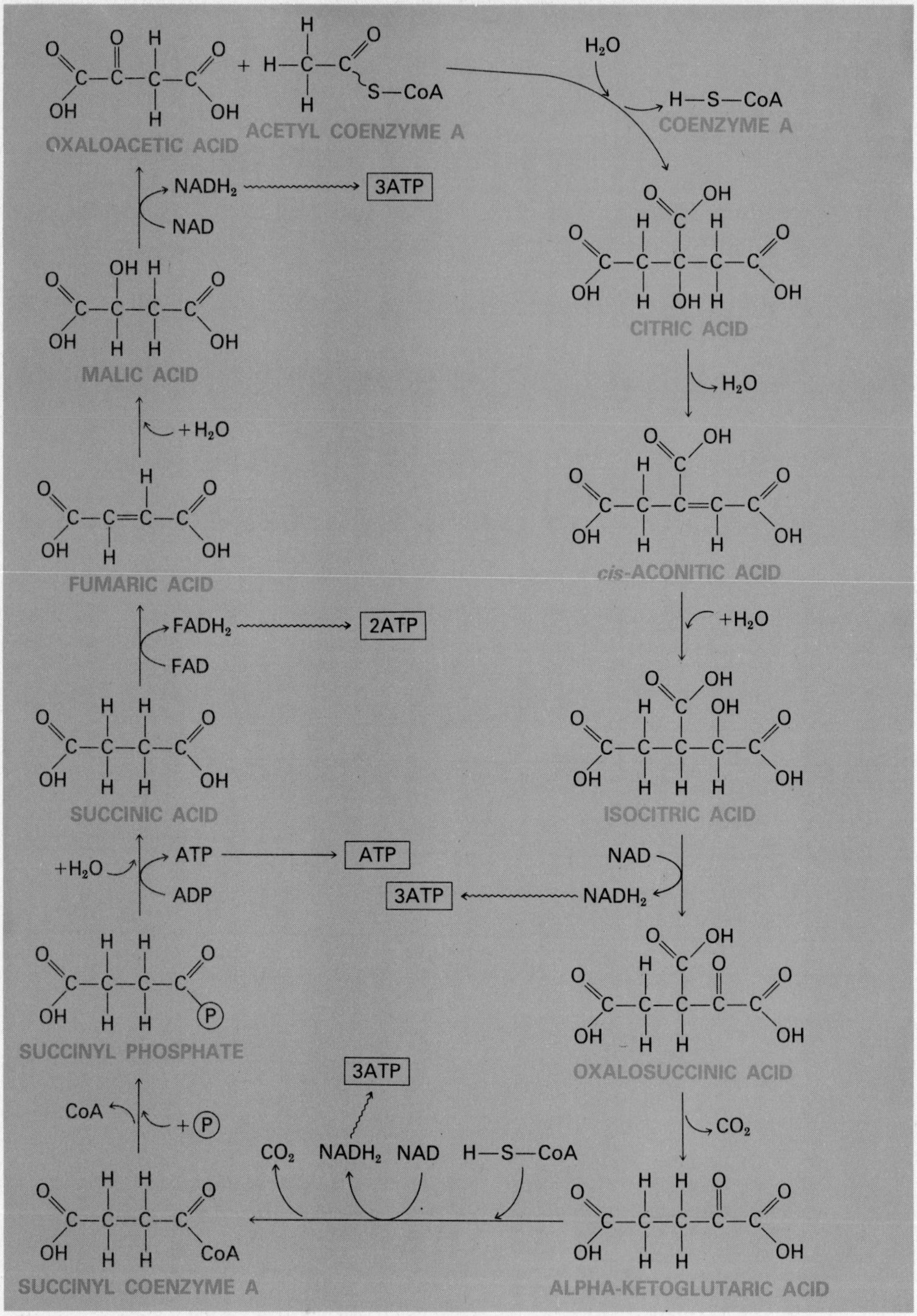

The Krebs citric acid cycle.

intricate series of additions, subtractions, and rearrangements, as shown in Figure 9.13, the six-carbon citric acid is converted into the four-carbon oxaloacetic acid. In the process two molecules of carbon dioxide are given off, one molecule of ATP is generated, and three pairs of hydrogens are picked up by NAD to produce three molecules of $NADH_2$. Each molecule of $NADH_2$ can pass through the series of cytochromes to generate three molecules of ATP. In addition, one further pair of hydrogens is picked up by the hydrogen-acceptor molecule, flavin adenine dinucleotide, or FAD. FAD is a derivative of still another vitamin of the B_2 complex, **riboflavin.** FAD behaves in very much the same way as NAD, but when the hydrogens from $FADH_2$ are passed through the cytochrome series, only two molecules of ATP are generated (see Figure 9.13). This whole series of reactions is known as the **Krebs citric acid cycle.** It is named after the man who first determined all its component reactions, Dr. Hans Krebs.

The most important aspect of the Krebs citric acid cycle for the cells that use it is that each acetyl molecule is eventually broken down into carbon dioxide and water, which generates 12 molecules of ATP. One molecule of glucose can produce 2 acetyl molecules. This means that one molecule of glucose results in the completion of two laps of Krebs's cycle, generating 24 molecules of ATP. Together with the 14 molecules of ATP gained by converting one molecule of glucose into 2 molecules of acetyl CoA, complete aerobic respiration of a molecule of glucose results in the net production of 38 molecules of ATP (see Figure 9.14).

THE SIGNIFICANCE OF ENERGY EXTRACTION

The first thing that you should notice is that aerobic respiration gives a much greater yield of energy from a molecule of glucose than the anaerobic processes of glycolysis and fermentation. The anaerobic processes can only give a net yield of 2 molecules of ATP, while aerobic respiration gives a net yield of 38 molecules of ATP. This makes aerobic respiration approximately nineteen times more efficient than anaerobic energy extraction. Thus from a given amount of glucose an aerobic organism can do about nineteen times more work than an anaerobic organism. Alternatively, an anaerobic organism must be supplied with nineteen times as much glucose in order to accomplish the same amount of work as an aerobic organism.

What kinds of work are we talking about? You will notice in the various figures illustrating the component parts of respiration that each specific reaction requires the presence of a different specific enzyme. These enzymes are all protein molecules that must be synthesized by the organism. The synthesis of all those proteins requires the expendi-

Figure 9.14

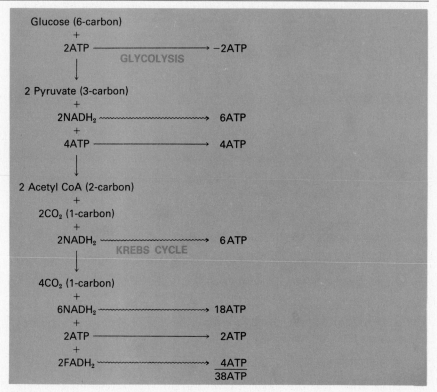

Summary diagram of aerobic respiration.

ture of energy—the use of molecules of ATP. The movement of molecules from one location to another may also require energy. For example, the $NADH_2$ produced by glycolysis occurs in the cytoplasm of the cell. The cytochrome molecules that can process the hydrogens to generate more ATP are located within the mitochondrion. It requires energy to move the $NADH_2$ through the mitochondrial membrane so that the full energy yield may be obtained. This may be just another molecular-energetic example of the old maxim that it takes money to make money, although in this case it takes energy to obtain more energy.

One of the most common difficulties experienced by most students at this stage of their introduction to respiration is in trying to comprehend where all the various molecules that have been introduced have come from. This concern is a legitimate one, and its answer tends to reinforce the importance of respiration and the whole subject of energetics. Ultimately the molecules being respired must be produced by photosynthesis. Even in the case of a strictly animal type of molecule, that compound was produced by using the energy that that animal (or some other animal) obtained by respiring a product of the photosynthetic process. All the other molecules involved in the process must either be produced by the organism itself or obtained from the outside. To some extent, some of these materials are supplied to a new cell by its predecessors, but after that the cell is on its own.

For single-celled organisms that live in an appropriate environment, obtaining some of the necessary molecules such as oxygen or water poses no real energetic problems. These molecules may just diffuse from the environment, where they are present in high concentrations, into the cell. For multicellular organisms such as ourselves, however, obtaining enough of even relatively abundant materials such as water and oxygen requires energy.

Most of the cells of a human being are far removed from the availability of the molecular oxygen in the atmosphere. As we discussed earlier, humans breathe, bringing large masses of oxygen-rich air into the lungs. This mass movement of air is accomplished by the contraction of appropriate muscles. Muscular contraction requires molecules of ATP. The building of the contractile proteins that are muscle also requires ATP. In the lungs the oxygen diffuses into the blood, more specifically, into the red blood cells, which contain huge volumes of hemoglobin. Hemoglobin is, as we have already seen, a complex protein molecule with a high affinity for oxygen. ATP molecules are required to synthesize hemoglobin. Red blood cells are carried to the vicinity of all living cells in the body by the moving stream of blood through the blood vessels. This moving stream of blood is caused by the rhythmic pumping of a muscular heart. Again, the contraction of heart muscles is powered by molecules of ATP.

Many organic molecules, such as NAD, FAD, ADP, coenzyme A, oxaloacetic acid, and a multitude of enzymes, are needed to complete respiration. Most of these substances are synthesized by the organism for use in respiration. All these molecule-building activities require ATP. Some organisms, however, are incapable of completely constructing these molecules. NAD, for example, cannot be completely synthesized by human beings. For this reason it must be taken into the body preassembled by some other organism, as part of the daily intake of food substances. Taking in food requires converting ATP molecules into ADP and phosphate. Most of the substances we call vitamins are molecules that function like NAD or coenzyme A but cannot be synthesized by the human body. Therefore, they must be ingested by humans after being fully formed by some other organism that *is* capable of synthesizing them.

For the sake of simplicity, the process of respiration has been shown as starting with the relatively simple carbohydrate molecule glucose. It was also stated that glucose was the traditional end point of photosynthesis. Most photosynthetic organisms are capable of converting glucose into a wide variety of other carbohydrates. Most organisms are also able to interconvert, with varying degrees of facility, carbohydrates, amino acids, and the different components of lipids. These interconversions usually require ATP, some specific enzymes, and occasionally some additional compounds. For example, in converting carbohydrates to amino acids, an appropriate source of nitrogen is required. This feature of interconvertability means that organisms are usually able to release the energy trapped in any complex organic molecule through the process of respiration.

Summary

1 / *Living things are complexly organized entities. Complexity requires energy for construction and maintenance.*

2 / *The breakdown of large organic molecules releases energy and materials used to produce and maintain living things.*

3 / *The ultimate producers of organic molecules are green, photosynthetic organisms.*

4 / *Photosynthetic organisms use the energy in sunlight to combine carbon dioxide and water into organic compounds (most notably carbohydrates) and release oxygen and water in the process.*

5 / *Under anaerobic conditions many organisms release some energy from organic compounds through the linked processes of glycolysis and fermentation.*

6 / *Cellular respiration is an intracellular process whereby some organisms obtain energy from organic compounds by using molecular oxygen.*

7 / *Breathing is a mass movement of gases in and out of specialized organs of the body.*

8 / *Gas exchange is the reciprocal diffusion of oxygen and carbon dioxide across membranes, most usually oxygen moving in and carbon dioxide moving out of the body.*

9 / *The most common form of energy involved in living functions is the conversion of a molecule of ATP into a molecule of ADP and phosphate. ATP is known as the universal energy currency.*

10 / *Glycolysis converts a molecule of glucose into two molecules of pyruvate. In the process it produces two molecules of ATP and reduces two molecules of NAD.*

11 / *In fermentation, reduced NAD ($NADH_2$) and pyruvate react to produce NAD and either lactate or carbon dioxide and ethanol.*

12 / *Fermentation can only occur in the absence of oxygen—it is an anaerobic process.*

13 / *Electron transport chains can convert high-energy, reduced NAD ($NADH_2$)*

into low-energy NAD by adding the two hydrogens to oxygen, producing water. The energy thus released can be used to generate three molecules of ATP from three molecules of ADP and three molecules of phosphate.

14 / Without $NADH_2$, pyruvate can be reacted with coenzyme A to produce acetyl CoA.

15 / The Krebs cycle is a long circular series of reactions by which the remaining carbon skeleton of pyruvate is completely dismembered for a net production of 12 molecules of ATP.

16 / Aerobic respiration releases almost twenty times as much energy in the form of ATP molecules as the anaerobic processes of glycolysis and fermentation.

17 / Vitamins are molecules that are essential to many living processes but that the organism is not capable of synthesizing itself. It is therefore necessary that they be part of the normal diet. Some vitamins are required to complete cellular respiration.

Questions to Think About

1 / How many different ways can you think of to trap energy in such a way that it can be used at some later time? How many of these ways are directly dependent upon photosynthesis?

2 / Under what kinds of conditions will the greatest amounts of photosynthesis take place?

3 / What effects do increasing urbanization have on the total energy available to living things?

4 / How many different sources of energy can you think of that can be used to accomplish work? How many of these depend on the existence of living organisms?

5 / If both aerobic respiration and burning will give the same energy yield from a given amount of organic compounds, which one is better, and why?

6 / Most people recognize that a varied diet makes eating more interesting. Is it also healthier. Why should this be so?

10

Ecology

Objectives

The student should be able to:

1 / explain the idea of trophic levels and describe the kinds of organisms that can be found in each of the different levels.

2 / describe what is meant by food chains and food webs.

3 / explain why most communities exhibit pyramids of numbers, biomass, and energy.

4 / describe the first and second laws of thermodynamics.

5 / give a clear example of the second law of thermodynamics.

6 / explain why natural communities seldom have organisms occupying a fifth or higher trophic level.

7 / describe the major categories of interspecific interactions.

8 / explain what is meant by biogeochemical cycling and why it should exist.

9 / explain what is meant by community succession and how it functions.

10 / describe the three major properties of a population that contribute to the dynamics of its growth.

11 / demonstrate how a geometric progression works.

12 / explain why most populations exhibit a sigmoid growth curve.

13 / explain how the concept of environmental carrying capacity sets a limit to population size.

14 / describe the course of events that occurs in a population that exhibits a J-shaped growth curve.

15 / explain why so many acts in the biological world have multiple repercussions.

16 / describe the growth of the human population for the last 4000 years.

17 / suggest the most probable future for humanity, taking into consideration everything that you have ever learned.

Introduction

The study of photosynthesis and respiration has led to the conclusion that all organisms extract energy and raw materials from organic compounds such as carbohydrates, lipids, and proteins. While all organisms are composed of these chemical substances, the ultimate producers of these carbon-containing compounds are the photosynthetic organisms—the green plants, algae, and some bacteria. Because only the photosynthetic organisms can harness the energy of sunlight to produce organic comounds, all other organisms—the heterotrophic organisms—are dependent upon these producers.

The existence of one species of organism as a food resource for another species is just one example of the ways in which different species interact within an ecosystem. The ways in which species interact, and the effects produced by these interactions, are the chief considerations of the subdivision of biology known as **ecology**. The study of ecology has resulted in several significant insights about the nature of life and the world we inhabit. The most significant of these insights is that different living things require one another for their continued existence.

Trophic Levels

Among the diverse living forms on the earth today, it is possible to imagine huge groupings or categories based on the most usual source of energy input. These categories are usually called **trophic levels,** or feeding levels. At the first trophic level are all the **autotrophic,** or self-feeding, organisms. Such organisms can produce all their own food from inorganic starting materials by utilizing an abiotic (nonliving) energy source. This trophic level must include all the photosynthetic organisms and those relatively insignificant bacteria that are autotrophic without being photosynthetic. The most common name given to the organisms within this trophic level is **producer.** All other organisms are **consumers,** but there are different levels of consumers.

The second trophic level is composed of all those consumers that feed directly on the producers. These organisms are called the primary consumers. Because most of the producers are plantlike organisms, and because primary consumers feed directly on these producers, the primary consumers can be described most easily as vegetarians, or **herbivores.** Familiar examples of herbivorous organisms include cattle, deer, rodents, seed-eating birds, and many species of insects.

The third trophic level consists of those organisms that feed on the primary consumers. These secondary consumers generally eat animals that are herbivorous, meaning that they are meat eaters, or **carnivores.** Familiar examples of carnivorous organisms include wolves, lions, snakes, insect-eating birds, and spiders.

A fourth trophic level is composed of organisms that derive their energy from the respiration of material obtained from members of the third trophic level. These organisms, which feed on carnivores, are known as tertiary consumers, or secondary carnivores. Examples of secondary carnivores include dog and cat fleas, many species of hawks, large game fish (like bass), and killer whales.

You can probably imagine progressively higher trophic levels composed of species that feed on the organisms in the trophic level below. By really stretching your imagination, you could probably envision a fifth- or sixth-level carnivore that would occupy the seventh or eighth trophic level. However, in nature we seldom find trophic levels higher than the fifth or sixth.

The ultimate group of consumers is the one that feeds on all trophic levels indiscriminately, the decomposers. The decomposers are not terribly fussy about what it is they consume, as long as it is dead. The trophic level of the food source is irrelevant, and some impatient decomposers don't even insist that their food source be dead. The commonest examples of decomposers include the fungi and many kinds of bacteria.

The idea of trophic levels (see Figure 10.1) has some minor imperfections in it, as you can easily see by contemplating the sources of food consumed by people (see Table 10.1). Human beings are organisms that can be both herbivorous and carnivorous, a condition usually described as omnivorous. Our food intake could place us anywhere from

Figure 10.1

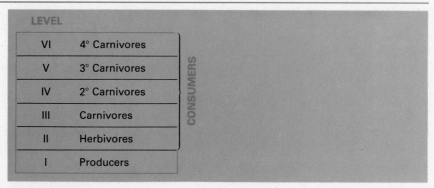

Trophic levels.

Table 10.1
Sources of Food

TROPHIC LEVEL	MAJOR FOOD ITEMS OF PEOPLE
Producers	Vegetables, fruits, nuts, grains, breads and pastas, sugars
Herbivores	Beef and veal, poultry and eggs, lamb and mutton, pork, milk, cheese, and other dairy products
Carnivores	Tuna, salmon, perch, bass, frogs, crab, shrimp, lobsters

the second trophic level up to the fifth or sixth. Many other species have similarly broad dietary intakes. As a sort of meaningless aside, what would you call an organism that eats decomposers, as we do when we consume mushrooms? Answer: a gourmet.

The immediate implication of trophic levels is that there should be chains of organisms proceeding from the lowest to the highest trophic levels, the organisms at each level being fed upon by some other organism in the next higher level (see Figure 10.2). In actuality, such **food chains** are difficult to find because there is a certain amount of flexibility in the feeding patterns of many consumers, as we have seen in humans. An examination of the feeding relationships among organisms in a natural community often reveals the presence of a **food web**. In a food web a particular consumer may feed on several different species that occupy several different trophic levels, and several other species of consumers may also be using these same sources of nutrition. At the same time there may be several different higher-level consumers trying to eat the original consumer. The feeding relationships within a natural community can be very complex, as shown in Figure 10.3.

TROPHIC PYRAMIDS

Even if trophic levels are only imperfect descriptions of the feeding relationships that exist among the organisms within a community, the basic idea can be useful in revealing certain generalizations about the living world. For example, one of the earliest questions to be asked about trophic levels concerned the relative number of individual organisms at each level.

Pyramid of Numbers If you work with an area of reasonable dimensions, it is possible to count all the individual organisms within that area and assign them to their appropriate trophic levels. This has been done several times and the results usually conform to what has been described as a **pyramid of numbers** (see Figure 10.4). The first trophic level, the producers, has a huge number of individuals. There are fewer herbivores, still fewer carnivores, and even lesser numbers of secondary carnivores. Higher levels of consumers are extremely rare, while the numbers of decomposers are variable. What makes this observation particularly interesting is that it is not unique to any particular kind of community. Many different kinds of communities—for example, grassland communities, forest communities, and many aquatic communities—all exhibit the same pyramid of numbers. There are usually

Figure 10.2

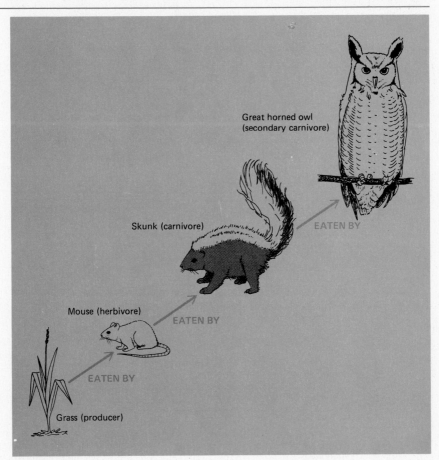

A food chain.

Figure 10.3

An example of the complex ways in which many species can interact within a food web. Most species have several different food items and are, in turn, the food items of several other species. [Adapted from R. F. Johnston, *Wilson Bulletin,* vol. 68, pp. 91–102, 1956. Copyright R. L. Smith, *Elements of Ecology and Field Biology,* New York, Harper & Row, 1977.]

Figure 10.4

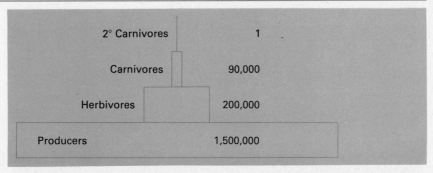

2° Carnivores	1
Carnivores	90,000
Herbivores	200,000
Producers	1,500,000

A pyramid of numbers. Numbers refer to individual organisms per 0.1 hectare, exclusive of microorganisms and soil animals. [Data taken from grassland studies by Evans and Cain, 1952, and Wolcott, 1937; as published in Eugene P. Odum, *Fundamentals of Ecology,* Philadelphia, Saunders, p. 80, 1971.]

more individuals in the next lower trophic level, and fewer individuals in the next higher trophic level.

One plausible explanation for the consistent pattern of the pyramid of numbers in many different communities is the observation that consumers are often larger than the organisms they consume. For example, rabbits are usually larger than the grasses they eat, and they, in turn, are eaten by larger carnivores, such as foxes. Perhaps the pyramid of numbers is just a reflection of this increase in size as we ascend the trophic levels.

The easiest way to test this particular hypothesis is to weigh the individuals in each trophic level. However, much of the weight of living organisms consists of stored water, which may be present in extremely variable amounts. We are not really interested in knowing how much water is being stored at the various trophic levels. What we do want to know is the accumulation of the organic compounds at the different trophic levels. For this reason we should measure the dry weight of the organism—its weight with all the stored water removed. This is called the organism's **biomass.**

Pyramid of Biomass   If we attempt to determine the distribution of biomass within a community, we will usually observe that a huge amount of biomass exists in the form of producers, a lesser amount is found in the herbivores, still less in the carnivores, and even lesser amounts in the higher levels of consumers. Again the amount of biomass tied up in decomposers is variable, but generally small. In brief, most communi-

Figure 10.5

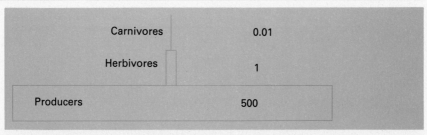

Carnivores	0.01
Herbivores	1
Producers	500

A pyramid of biomass. Numbers refer to grams of dry weight per square meter. [Data taken from Odum, 1957; as published in Eugene P. Odum, *Fundamentals of Ecology,* Philadelphia, Saunders, p. 80, 1971.]

ties exhibit what can be described as a **pyramid of biomass** as well as a pyramid of numbers (see Figure 10.5). Not only are there more individuals at the producer level, but there is a greater mass of living material at the producer level. Higher-level consumers not only are rare individuals but constitute a relatively insignificant portion of the total biomass of a community.

Pyramid of Energy   The consistent existence of a pyramid of biomass suggests that each successively higher trophic level has less material in the form of organic compounds to offer as food for the next higher level. Thus each successive trophic level must have less in the way of raw materials and energy to sustain living functions. It is possible to test this hypothesis by measuring the amount of energy trapped in the organic compounds of the different trophic levels in different communities. This, too, has been done and, not surprisingly, it has been found that all communities exhibit what can be called a **pyramid of energy** (see Figure 10.6). A large amount of energy is present at the producer level and progressively less energy is available in successively higher trophic levels.

THE SECOND LAW OF THERMODYNAMICS

Why should these patterns exist? In particular, why should the distribution of individuals, biomass, and energy in the trophic levels of a community be in the form of a pyramid? The most reasonable explanation for these patterns comes from the physical sciences in the form of the laws of **thermodynamics.** There are several laws of thermodynamics that have been of particular importance to chemists and physicists, but only the first two need concern us.

The first law of thermodynamics states that energy can be neither created nor destroyed but that it can be converted from one form to another. Thus the energy of sunlight can be converted into the energy of chemical bonds in the process of photosynthesis. Einstein's theory of **relativity** modified this law somewhat to show that matter and energy are interconvertible, but as the vast majority of living organisms are incapable of making use of this principle, we will ignore it.

The second law of thermodynamics states that when energy is converted from one form to another, the conversion is never 100 percent efficient. Some of the energy is always converted to a third form that is not usable within the system and thus is effectively lost.

It is important to see that the second law of thermodynamics does not contradict the first law. In energy conversion all the energy is still present; none of it has been destroyed. It's just that some of it is not present in any usable form. For example, we all know that electrical energy can be converted into light energy by using an appropriate bulb. However, the amount of energy in the form of light from the bulb is less than the amount of electrical energy that goes into the bulb. The difference exists in the form of heat energy also given off by the bulb. This heat performs no useful function within the system and is effectively unusable.

It is the second law of thermodynamics that appears to be most relevant to understanding the pyramidal structure of the trophic levels. For all intents and purposes, all living things are dependent upon green plants for their supply of usable energy. Green plants are dependent upon the energy of sunlight to trap their supply of usable energy.

Figure 10.6

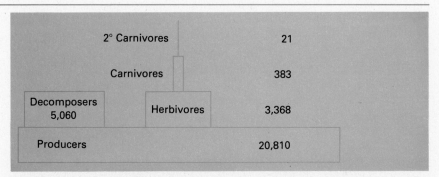

A pyramid of energy. Numbers refer to kilocalories per square meter per year. [Data taken from Odum, 1957; as published in Eugene P. Odum, *Fundamentals of Ecology*, Philadelphia, Saunders, p. 80, 1971.]

Just how much energy is available in the form of sunlight? Each year the earth's surface receives about 5.5×10^{23} **Calories** (abbreviated Cal) of solar radiation. That is approximately 100,000 Calories per square centimeter per year. Of this total amount approximately a third goes into pleasant, but organically useless, work such as evaporation of water and warming the land and water and air. This leaves approximately 67,000 Calories per square centimeter per year for activities such as photosynthesis.

Every year photosynthesis converts about 200 billion tons of carbon in the form of CO_2 into organic compounds. For those of you who are not impressed by the enormity of that number, human beings only manage to produce about 2 billion tons of goods in a year. Despite the magnitude of material turnover, the efficiency of energy conversion is dismally poor. The second law of thermodynamics is still in effect, and some of the energy being converted into chemical bonds winds up in some unusable form such as heat. The amount of energy actually trapped as photosynthetic products in a year's time is somewhere in the neighborhood of 33 Calories per square centimeter per year.

Knowing the amount of sunlight energy available for photosynthesis (67,000 Calories per square centimeter per year), and knowing the amount of energy trapped in photosynthetic products (33 Calories per square centimeter per year), we can calculate a rough approximation of the efficiency of photosynthesis: 33/67,000, or 1/2,000, or 0.05 percent. This efficiency figure is almost certainly not exactly correct in that much of the surface area of this planet that receives solar radiation does not have plant life to carry out photosynthesis, or does so for only part of the year. This could probably throw the estimation off by 10 or even 100 times. However, even if the estimate is off by 100 times, photosynthetic efficiency is only around 5 percent. Photosynthesis apparently wastes a tremendous amount of energy; most of the energy that reaches the planet and the plants on it does not get locked into photosynthetic products. Photosynthesis may not be very efficient, but it does trap an average of 33 Calories per square centimeter per year in organic compounds, and it is the only natural process that does so.

Does this mean that the 33 Calories per square centimeter per year trapped by photosynthesis are available to all the consumers in the world? Unfortunately it does not. Most of this trapped energy is used by the producers themselves in the process of staying alive. There are some differences of opinion over just how much energy plants use to stay alive, but most investigators will agree that an estimate of 20 percent of organic compounds available as biomass in the plant is probably generous. Thus plants use about 80 percent of the energy trapped by photosynthesis just to keep themselves alive. This leaves about 20 percent of the photosynthetic energy—about 6.6 Calories—available in the form of organic compounds that can support consumers.

Herbivores, the primary consumers, use much of the energy they take in to perform all the vital functions that constitute staying alive. Once again there are differences among the different species of herbivores, but an average efficiency of 10 percent seems reasonable. Thus about 90 percent of the energy intake is used to keep the herbivore alive, while the remaining 10 percent of the energy intake is stored as herbivore biomass. It is the herbivore biomass—about 0.66 Calorie per square centimeter per year—that is available to supply the energy needs of the higher trophic levels.

Carnivores, the secondary consumers, are about as energetically efficient as the herbivores. They require about 90 percent of their energy intake to keep themselves alive, while only about 10 percent of this energy intake gets converted into carnivore biomass. This means that approximately 0.06 Calorie is available as energy in organic compounds for secondary carnivores.

For each transition to a higher trophic level, there is only about 10 percent as much energy available. The remainder is converted to a form that is not usable by other organisms; usually this form is heat. The loss of available energy is accomplished by keeping the organisms at the given trophic level alive. This fact explains generally why there are fewer herbivores than producers; fewer carnivores than herbivores; fewer secondary carnivores than primary carnivores. It also places a practical limit to the height of trophic levels. The reason we seldom find communities with more than five or six trophic levels is that so little energy is available, or it is so widely dispersed, that organisms trying to feed in so high a trophic level are incapable of meeting their energetic needs.

The main idea that you should get from this information is that living things on this planet must work within the confines of an energy budget. Some amount of energy is given to the earth in the form of sunlight, and this energy flows through the various trophic levels in ever-diminishing amounts. Some of it becomes temporarily trapped as organic molecules in living material. Relatively small amounts of energy may be trapped for long periods of time as preserved organic compounds, such as coal or petroleum. But almost all this energy is dissipated into space by the members of each trophic level as unusable heat. During its flow through the trophic levels, the energy can be used to maintain the dynamic equilibrium that we call living.

The total energy budget for a trophic level is not necessarily completely available to any one particular species within that trophic level. As an example, consider a simple food chain consisting of grass, rabbit, and fox. Each of these species occupies a different trophic level. Some amount of energy is tied up in grass biomass, which can support rabbits, and some lesser amount of energy is tied up in rabbit biomass, which can support foxes. The amount of energy involved in rabbit biomass establishes the energy budget for the carnivore trophic level. All this energy is not available to foxes, however. Many organic molecules in rabbits, notably those involved in the structure of hair, bone, teeth, and claws, cannot be used by foxes because they are indigestible. This apparent waste of energy will usually be exploited by some other life form. The food webs of most ecosystems appear to be set up in such a way that the maximum amount of available energy gets used.

Direct Interactions Between Species

Many of the ways in which two species may interact are directly observable and describable. From an observational standpoint, three categories of interactions are apparent. These categories are determined by the effects on the two species involved in the interaction. Two species can interact in ways that are advantageous to one but disadvantageous to the other, disadvantageous to both, or advantageous to both. These three categories of interactions have been called **predation** or **parasitism**, **competition**, and **mutualism** or symbiosis, respectively.

PREDATION OR PARASITISM

When one organism eats or consumes another, it is generally considered advantageous for the eater and disadvantageous for the one being eaten. Two different names have been applied to this situation depending upon the particular circumstances. It is usually called predation when the eater, or predator, is larger than the organism that is eaten, the prey. Predators most commonly kill their prey quickly. The opposite extreme of this condition is called parasitism. In parasitism the organism doing the eating, called the parasite, is generally much smaller than the organism being eaten, called the host. As a consequence of this size differential, the consuming is done very slowly and frequently results in a very protracted relationship. In the best parasitic relationships the host is not killed by the parasite. As you can imagine, many intermediates between these two extreme positions can be found.

A predatory or parasitic relationship presents a set of environmental conditions to which each of the participants must adapt if their lineages are to survive. For example, prey species have frequently evolved characteristics that make them more difficult to consume by the usual predators. Such characteristics include spines and hairs on many plants and insects, noxious sprays produced by some insects, storage of certain metabolic products that make the prey unpalatable, cryptic colorations, speed and maneuverability, and a multitude of behavioral patterns. One of the most common mechanisms that prey species use to maximize their survival is the maintenance of a very high rate of reproduction. The basic strategy seems to be that if a large number of individuals from a population are going to be lost to predation, it is best in the long run to produce even more offspring. In this way the chances are maximized that enough offspring will survive to maintain the continuity of the species.

Predators and parasites must evolve more and more characteristics that maximize their ability to find and consume their prey/hosts without overexploiting their food resource. A very effective predator or parasite is a catastrophe, because it not only eliminates its prey/host but at the same time eliminates itself. It is difficult for a predator to survive very long after it has killed off all its food supply. Internal parasites, in particular, have evolved a remarkable array of characteristics designed to maintain the continuity of their species.

COMPETITION

When two or more species are attempting to use the same resource at the same time, they are said to be in competition for that resource. An obvious example of competition occurs in gardens across the country every year as people try to utilize vegetable resources that are highly preferred food items of birds, rabbits, and a multitude of insect species. Competition tends to be disadvantageous for both species in that neither can do as well in the presence of the other as either species could do in the absence of the other. Obviously, the more intense the competition, the more disadvantageous the situation is. Careful studies of competitive relationships in both the laboratory and nature have revealed an overwhelming tendency to minimize the extent of competition between species. Usually under a given set of circumstances, one of the two species can outcompete the other, leading to the eventual elimination of the less successful competitor. In nature, where circumstances tend to be very variable, individuals of the two species will tend to reproduce most effectively when the overlap in their resource utilization is minimized (see Figure 10.7). This results in a gradual separation of their specific preferences and a lowering of the competition.

MUTUALISM

Mutualistic relationships are those in which both species involved benefit from the relationship to such an extent that the relationship is obligatory. Such relationships are sometimes referred to as symbiotic. An excellent example of a complex mutualistic relationship is found

Figure 10.7

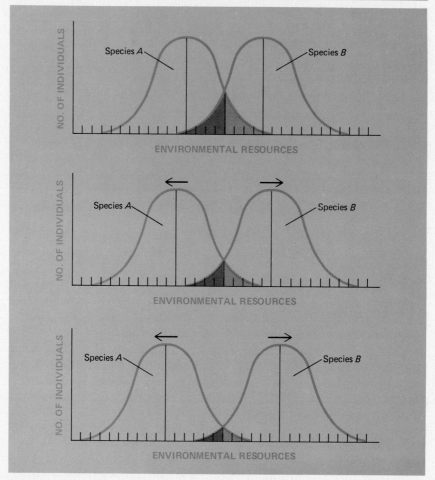

Evolving to minimize competition. Competition minimizes survival in areas where resources overlap.

between cattle and several species of microorganisms. Cattle eat grasses and other plants that have a high content of cellulose, a polymer of glucose. However, like most vertebrates, cattle cannot produce enzymes for digesting cellulose into glucose and thus are unable to use all the energy locked up in this complex carbohydrate molecule. Many microorganisms can produce the enzymes for digesting cellulose, but some of them require a warm, dark, moist place in which to do so. By inhabiting a special chamber of the cow's stomach, microorganisms such as *Entodinium, Polyplastron, Isotricha,* and *Dasytricha* are provided with appropriate conditions and an abundant supply of cellulose. The microorganisms break down far more cellulose than required for their own maintenance. The excess is absorbed by the cow. This arrangement is extremely beneficial for each of the species involved. In fact, the interdependence between these species is so intense that their separation would result in the death of all.

The high level of intricacy involved in any mutualistic relationship inspires wonder in the human mind. How could such a magnificent partnership have been brought into existence? At this stage in our knowledge of biology, the answer to such a question must, of necessity, be somewhat speculative. Many mutualistic relationships appear to be derivable from associations that originally may have been parasitic or competitive. If we assume that some remote ancestors of cattle may have been much more omnivorous than cattle of today, it is possible that they could have become infested by any number of microscopic organisms. Many of these may have resided within the digestive tract, where they had ready access to all the food taken in by the ancestral cow. Such a relationship would not have been advantageous to the host unless sufficient additional foodstuffs were released from ingested plant products to offset that consumed by the microorganisms. Once these circumstances prevailed, natural selection would operate on both parties to change in ways that would maximize their abilities to function cooperatively.

I have used this particular case because I think it does an unusually good job of illustrating that the three categories of interspecific interactions are not as clear-cut as it would first appear. All three categories can blur together. Most organisms found in the digestive system of another organism are almost automatically called parasites. As we have seen, some of them may be mutualistic rather than parasitic. Furthermore, many inhabitants of digestive systems consume only the food resources of their host rather than the host itself. Strictly speaking, it would be more accurate to refer to this as a competitive relationship, albeit a very intimate one. The dynamic conditions of life often resist the simple restrictions of our vocabulary.

Ecosystem Interactions

The kinds of interactions with which we have dealt so far are direct and straightforward. They have all been concerned with the procurement of energy. With the exception of the producers, all organisms are directly dependent upon the existence of some other organism as a source of energy in the form of organic compounds. These consumers, including the decomposers, depend on these same energy suppliers to provide the basic materials used in making more consumers. This direct dependency of the consumer results in a very fundamental kind of interaction between species. Consumer species must consume some other species.

Matter, the physical stuff of which this world is composed, is anything that occupies space and has mass—which is the same as saying that it weighs something. According to the physical sciences, matter can neither be created nor destroyed. In this respect matter has some of the same characteristics as energy. What this says for us is that for all practical purposes the amounts of the 92 elemental forms of matter present on the earth today are roughly the same as they have been for the duration of the earth's existence. To put it another way, all the atoms involved in the origin of the primitive earth are still around and *only* those original atoms are around. There are a few exceptions to that generalization, such as the atoms added to the earth by impacting meteorites or the atoms lost in space as a result of our explorations of the solar system, but these kinds of perturbations have not significantly altered the composition of the planet. For the most part we have the same number of oxygen atoms, carbon atoms, iron atoms, aluminum atoms, and so forth, as we have ever had.

If the number of atoms, the basic building blocks of matter, is essentially constant, but if, as we know, the appearance of the world is constantly changing, then it should be readily apparent that the building blocks are not static. Atoms must get shifted around, reorganized, rearranged, recombined, and recycled. This idea of moving and recycling matter is a crucial concept in the study of interacting organisms. Organisms interact with their nonliving surroundings as well as with other species, forming an abstract entity called the ecosystem. Ecosystems usually function in a way that results in the cycling of matter.

BIOGEOCHEMICAL CYCLES

The cycling of matter through an ecosystem can be demonstrated most easily by considering the element carbon. Carbon is found in the nonliving atmosphere in the form of carbon dioxide. Plants and other photosynthetic organisms take up carbon dioxide from the atmosphere and incorporate it into the structures of organic compounds. Eventually, either by the organism that synthesized the organic compound or by some other consumer, the organic compounds will be respired and the carbon will be released into the world as carbon dioxide; and the cycle is complete. Notice that producers provide consumers with carbon in an appropriate form, while at the same time consumers are providing producers with a supply of carbon that is appropriate to them (see Figure 10.8).

Figure 10.8

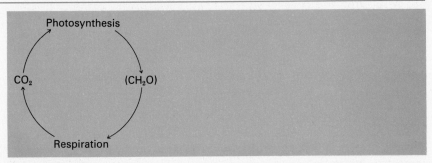

The carbon cycle.

A more complex kind of cycling occurs with respect to the element oxygen. Aerobic organisms use molecular oxygen removed from the atmosphere, or dissolved in water, to accept the hydrogens removed from organic compounds in respiration. The water so formed is released into the environment, where it may do any one of several things. As liquid water it may sink into the ground, forming part of a subterranean aquifer, or it may flow across the surface joining the system of streams, rivers, and oceans. Eventually it will probably be evaporated as water vapor into the atmosphere and precipitate in the form of rain, snow, sleet, or hail. It is also possible that an individual water molecule may be picked up by a photosynthetic organism, either as liquid water from the ground or from an aquatic system or as water vapor in the atmosphere. As you will recall from the discussion of photosynthesis in Chapter 9, the oxygen of water used in photosynthesis is released as molecular oxygen, and the cycle is completed (see Figure 10.9).

Because there is usually a complex interaction among the land, the waters, the atmosphere, and living organisms in the recycling of matter, these complex processes are usually called **biogeochemical cycles**. As you can see from the two examples already given, the dependencies among organisms are not quite as one-sided as it first appears. Not only do consumers require the existence of producers, but the continued existence of the producers is aided significantly by the existence of the consumers. The decomposers play an absolutely crucial role in the functioning of ecosystems by guaranteeing that the organic molecules in dead organisms are not permanently removed from the biogeochemical cycles.

Some additional aspects of interspecific interdependencies can be revealed by examining an even more complex biogeochemical cycle,

Figure 10.9

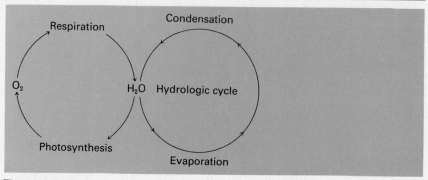

The oxygen cycle.

such as the one for nitrogen (see Figure 10.10). Nitrogen is an important element for living things; it is found in NAD, ATP, nucleic acids, and proteins. All these nitrogen-containing compounds are invariably associated with living materials. Proteins, for example, are composed of amino acids, which contain one or more amine groups ($-NH_2$). However, many consumers are incapable of synthesizing the amine group, and thus they cannot make amino acids. This means that they must take in preformed amine groups from those organisms that are capable of synthesizing them. Plants are generally capable of synthesiz-

Figure 10.10

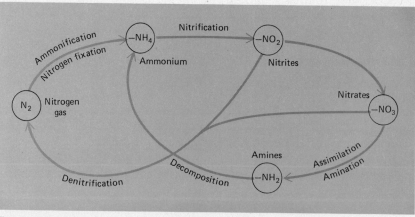

The nitrogen cycle.

ing amines from two different groups of nitrogen compounds called **nitrates** (—NO_3) and **nitrites** (—NO_2). Nitrites and nitrates are produced by an obscure group of microorganisms, called **nitrification** bacteria, from the nitrogen compound **ammonia** (NH_3). Ammonia and ammonium ions (—NH_4) can be produced either through the decomposition of amines by decomposers or through the **ammonification**—also called nitrogen fixation—of nitrogen gas (N_2) by still another group of microorganisms, called nitrogen-fixing bacteria. Yet another group of microorganisms, called **denitrifying** bacteria, can convert the nitrites and nitrates into elemental nitrogen gas.

The whole point of this rather involved illustration is that there are indirect dependencies as well as direct ones. Many organisms, such as human beings, require preformed amines and are dependent, therefore, upon their food organisms to provide them. But plants, as an example of the kind of organism upon which we are directly dependent, require the presence of nitrification bacteria to produce the nitrites and nitrates used to produce the amines, and the nitrification bacteria require nitrogen-fixing bacteria or decomposers or both to produce the ammonia from which they synthesize the nitrites and nitrates. Even though we do not consume them directly, we are just as dependent on the proper functioning of the nitrogen-fixing bacteria, the decomposers, the denitrifying bacteria, and the nitrifying bacteria as we are on the plants and animals we do consume. Without all those little nitrogen converters, our food sources would cease to exist, and so would we.

There are, or course, many other elements important to the continued functioning of living organisms, and each one has a different pattern for its biogeochemical cycling. Each cycle involves a different array of organisms that perform a vital transformation of chemical composition. The success or failure of any living species is almost certainly dependent upon the continued existence of a whole multitude of other living species with which it may have only remote connections.

COMMUNITY SUCCESSION

Many times the interdependence and interactions of different species take on some characteristics that, at least superficially, may appear to be surprising. For example, there's a well-known observation that the species composition of a community has a strong tendency to change through time. This phenomenon is called community **succession.** If you start with a particular piece of real estate lacking any organisms, many different plant and animal associations will appear and replace one another in a predictable sequence over a reasonably predictable duration of time until a stable, self-perpetuating **climax** community is achieved. The particular communities that occur and their sequence will depend upon the kind of soil and the general climatic conditions in the area of study.

A number of fascinating observations and generalizations have been made from studies of community succession, but the most startling is this: except for the species of a climax community, most organisms have a strong tendency to modify their surroundings in such a way that their continued existence becomes less likely. Let me support that generalization with a specific example.

In many parts of the United States where deciduous hardwood forests constitute the climax community, one of the earlier successional stages is a community dominated by pine trees. Pines do very well in poor, hard soils and with a great deal of exposure and limited moisture. Pine seeds germinate well and produce many seedlings under these conditions. Weedy old fields are among the most common places to find young pine trees. As time passes, and the young pine trees grow older and larger, they cause several changes in their immediate surroundings. Their root systems grow into the soil, loosening it and allowing the freer passage of water and air; pine needles cover the ground under the tree, providing organic material, water-holding capacity, and a certain amount of thermal insulation of the soil; and the tree itself provides shade, which cuts down on the amount of light and heat at the ground level. All these conditions are less than optimum for the production of new pine trees. It is difficult to find pine seedlings growing under a stand of mature pine trees, because conditions under mature pine trees are all wrong for the establishment of a new generation of pines. These new conditions under older pine trees are ideal for the germination and growth of seeds produced by trees with a greater tolerance of shade, the hardwoods, such as maples, oaks, and hickories. As the older pine trees die, they are replaced by the hardwood trees growing underneath them. As the composition of the forest changes from one composed of pines to one composed of hardwoods, the consumers change from those that feed on pines and their associates to those that feed on hardwoods and their associates.

Population Dynamics

On several different occasions within this book we have given consideration to different aspects of change in living systems. Growth and differentiation are examples of change within an organism. Evolution is an example of change by populations. Community succession is an example of change by communities. Populations can also exhibit a kind of change that is similar to that of individual organisms, namely, growth.

But population growth can differ from that of individual organisms because population growth can be negative. A population is capable of decreasing in size, or shrinking, as happens to a population of pines when they are replaced by hardwoods.

The kinds of interactions among species and their surroundings that we have been discussing in this chapter must have some effects on the dynamics of population growth. In addition, there are also characteristics, intrinsic to the population itself, that affect how that population grows. The most important characteristics of the population that determine its growth characteristics are the **birth rate,** the **death rate,** and **longevity.** When considering small geographic subdivisions of a large population, the rates at which individuals enter and leave the population by means of **immigration** and **emigration** may be important also.

The birth rate is usually defined as the number of new individuals added to the population by some standard number of individuals already in the population within some standard unit of time. For example, we might say that the birth rate is 15.6 live births of human beings in the United States per 1,000 people per year. It should be stressed that this is a statement that is a statistical summary. It does not mean that if you locked 1,000 people up together for a year that they would therefore produce 15.6 new offspring. However, on the average, during the year 1972, for every 10,000 people who lived in this country on January 1, there were 156 new offspring added by December 31. Notice that 156/10,000 is the same as 15.6/1,000. This same relationship can be expressed as 1.56 percent per year.

Death rates are defined in essentially the same way as birth rates, except that they record the proportion of individual organisms in the population that die within some standard unit of time. For example, the death rate for the same human population referred to above was 9.4 per 1,000 people per year. This means that during that particular year 94 people out of every 10,000 died. This death rate can also be expressed as 0.94 percent per year.

Longevity is a little more difficult to define, but it essentially has to do with the probability of an individual organism living to any particular age. This rate of survival will affect the age distribution of the population, which will, in turn, have an effect on both the birth rate and the death rate.

The interaction of these three population traits determines whether the population increases, decreases, or remains constant in size. If the birth rate exceeds the death rate, and the longevity remains constant, as in the examples given above, then the population will increase in size at a rate of 0.62 percent per year. If the death rate should exceed the birth rate, then the population must decrease in size at a rate equal to the difference between them. Obviously, if the birth rate equals the death rate, we would expect the population size to remain constant. A change in longevity of the population can cause a drastic alteration in the expected patterns of population growth. For example, a population with high longevity and a high representation of the older age groups will have a very different growth pattern from a population with lower longevity.

EXPONENTIAL GROWTH

The important thing to keep in mind is that the growth patterns of populations are rates of increase or decrease, which function in exactly the same way as compound interest rates on loans or savings accounts. Perhaps the most powerful illustration of the wonders of **geometric progressions** or **exponential growth**—or even compound interest—involves the mythical story of the hungry but clever young man. This bright young fellow offered to work for a local shopkeeper if the man would but double his salary every day. The first day's wages were to be one cent; two cents for the second day; four cents for the third day; and so on. It's hard to find shopkeepers that gullible any more. As can be seen in Table 10.2, the young man was only earning $5.12 per day on the tenth day of work. By day 15 he was making $163.84 per day, and by day 20 he was collecting $5,242.88 for a day's work. Had the shopkeeper been able to stay in business, the rich young man's salary on day 30 would have been $5,368,709.12.

Exponential growth—an increase by some constant rate—has a way of producing very large numbers very quickly if either the initial sum or the rate of increase is fairly large. Table 10.3 demonstrates how a large initial sum of $10,000 can be increased significantly by accumulating 5 percent interest compounded annually. In 20 years the original sum of $10,000 has generated $16,000 of interest. The original $10,000 can double itself in 15 years. Even more striking is what can happen to the same $10,000 kept for 20 years at a 5 percent annual rate compounded monthly. In this latter case the earned interest is added to the original sum every month and serves as the basis for the next month's calculation of interest. This results in a slightly greater rate of growth, meaning a larger total at the end of 20 years and a faster doubling of the initial sum.

What all this means is that populations of organisms are potentially capable of changing size and that characteristics of the population govern the rate at which the change in size occurs. However, rates of change say nothing about the absolute size that the population will achieve. Rates are simply a measure of how quickly or how slowly a

Table 10.2

Growth of Salary by a Factor of Two for 30 Days.

DAY NUMBER	DAILY SALARY
1	$ 0.01
2	0.02
3	0.04
4	0.08
5	0.16
6	0.32
7	0.64
8	1.28
9	2.56
10	5.12
11	10.24
12	20.48
13	40.96
14	81.92
15	163.84
16	327.68
17	655.36
18	1,310.72
19	2,621.44
20	5,242.88
21	10,485.76
22	20,971.52
23	41,943.04
24	83,886.08
25	167,772.16
26	335,544.32
27	671,088.64
28	1,342,177.28
29	2,684,354.56
30	5,368,709.12

change in population size can be achieved. To predict the eventual size of a population in absolute numbers, one must take into consideration the interactions among the organisms and their surroundings. All organisms, without exception, are dependent upon their surroundings—the environment—to provide those criteria necessary to sustain life.

S-SHAPED GROWTH CURVES

It is possible to see the effects of the interaction of population and environment on the growth of the population by studying simple orga-

Table 10.3
$10,000 Savings Account at 5%

YEAR	COMPOUNDED ANNUALLY	COMPOUNDED MONTHLY
0	$10,000.00	$10,000.00
1	10,500.00	10,511.61
2	11,025.00	11.049.39
3	11,576.25	11.614.69
4	12,155.06	12,208.91
5	12,762.81	12,833.53
6	13,400.95	13,490.10
7	14,071.00	14,180.27
8	14,774.55	14,905.75
9	15,513.27	15,668.34
10	16,288.94	16,469.95
11	17,103.38	17,312.57
12	17,958.55	18,198.29
13	18,856.48	19,129.34
14	19,799.30	20,108.01
15	20,789.27	21,136.76
16	21,828.73	22,218.14
17	22,920.17	23,354.84
18	24,066.18	24,549.69
19	25,269.49	25,805.68
20	26,532.96	27,125.93

nisms in relatively simple environments. If we inoculate a very small number of yeast cells into a bottle of growth-promoting materials and then count the cells over a period of time, we find that the total number of cells increases at some geometric rate for a while; then the rate of increase declines, until the population achieves some constant size for an extended period of time. The population size seems to reach some level above which it does not expand (see Figure 10.11).

The reason the population stops growing can be found in two changes in the population's environment. The first of these changes is an increase of waste products released by the organisms in the population. These materials are frequently detrimental to continued growth and reproduction of the organisms. The second change is a decrease in the amount of resources available to nurture continued growth and reproduction. In effect, it becomes impossible to obtain the materials to produce a new individual until an old one dies and releases those substances it has tied up.

In the early stages of this set of observations, the birth rate obviously exceeds the death rate because the population is growing. But as time goes on the birth rate exceeds the death rate by less and less until the birth rate and the death rate are equal and the population size is stable. What is interesting about these facts is the conclusion that the population and the environment are interacting to change the characteristics of each. As the population grows, it alters the environment. The changed environment alters the birth and death rates, which are characteristics of the population.

If such observations about population growth were restricted to microorganisms growing in bottles, we would probably be correct in considering them of questionable significance. However, the same pattern of growth has been found repeatedly in most kinds of populations that have been examined. A graph of population growth usually appears as an S-shaped curve, or **sigmoid** curve, as shown in Figure 10.11. For example, Figure 10.12 shows a graph of population growth for a population of domestic sheep introduced to the island of Tasmania.

LIMITS

When dealing with wild populations of a relatively complex sort, it is often noticed that the plateau reached by population growth is not per-

Figure 10.11

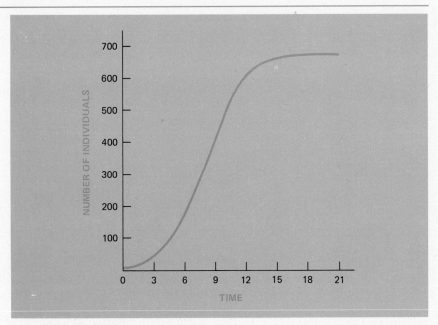

A sigmoid growth curve.

Figure 10.12

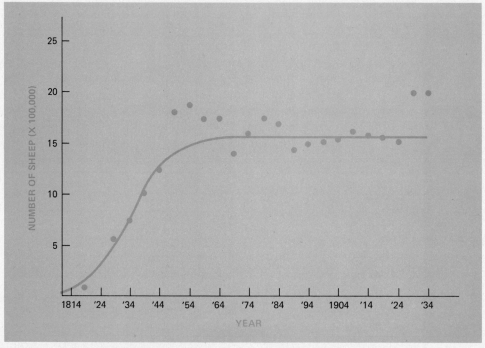

Population growth curve for sheep in Tasmania.

fectly flat but fluctuates irregularly up and down around some imaginary average. As we learn more about the details of the lives of the organisms in question, it becomes apparent that the fluctuations in population size are often due to variations in climate, which alter the conditions of the environment. Because the environment supplies all the resources needed to maintain a population, it naturally follows that the total amount of those resources available within the environment establishes the total capacity of that environment to support a given population. This **carrying capacity** of the environment for a population is the limit above which the population does not grow. Under different physical conditions an environment's carrying capacity may set different limits to population growth.

The idea of limits is an important one in the study of population growth. As previously noted, almost any population of organisms requires an array of materials to support continued growth. The amounts of the different materials needed vary under different environmental conditions and also vary among the different kinds of species. Any required substance can act as a limit to population growth, once its availability has been exhausted. This fundamental observation is known as **Liebig's law** of the minimum. It states that whatever substance is present in the environment in amounts closest to that needed by the population will be the substance establishing the limit to the population's growth.

As the simplest kind of example, suppose you have a population of organisms that requires only oxygen, carbon dioxide, and water. These organisms are exposed to the atmosphere, so they have ready access to an abundant supply of both carbon dioxide and oxygen. But suppose also that this is a desert situation and water is not very abundant. The population's requirement for water and the relative availability of water will limit the size to which the population can grow. It does not matter how much oxygen or carbon dioxide are available, because those resources are unusable without more water. Supplying more water to this environment would effectively raise the carrying capacity; it would raise the limit to the population's growth.

A similar law of tolerances has been expressed by an investigator named Shelford. Shelford found that for many measurable environmental factors there is a range of conditions within which the population could survive and grow. However, when conditions exceed this acceptable range in either direction, the population either ceases to grow, ceases to live, or both. For example, largemouth bass (a species of freshwater game fish) can survive and reproduce within a temperature range of $19°-27°C$. They can survive, but not reproduce, over an even wider range of temperatures ($5.5°-36.4°C$). If the temperature drops below $5.5°C$ ($42°F$) or rises above $36.4°C$ ($97.5°F$), then all individual bass die and the population size declines to zero.

The availability of required materials and the existence of ranges of tolerance for many conditions define the limits of population growth and the carrying capacity of the environment. Because the vast majority of living populations appear to remain reasonably constant in size over an extended period of observation, it seems most reasonable to conclude that they are most probably living at the carrying capacity of their environments. They have reached the limits of the environment's ability to support them. There are, however, some notable exceptions to the normal constancy of population size.

An increasing number of populations in today's world are declining in size or remaining stable in size at a critically low level. These populations either are associated with a particular environment that is being eliminated or are on the losing end of intense competition with some other species. The most probable outcome for such populations, based on past experience, is extinction.

J-SHAPED GROWTH CURVES

There is another group of species that do not exhibit a normal S-shaped growth curve. These species—the most notable examples being lemmings and thrips—manifest what has been called a J-shaped growth curve. Populations that grow in this way (see Figure 10.13) increase geometrically without coming into any apparent equilibrium with the carrying capacity of the environment. Such an increase in numbers is usually described as explosive. The environments of these organisms have the same kinds of limits as those present for populations exhibiting S-shaped growth curves, but populations exhibiting J-shaped curves seem to ignore the limits and expand beyond the carrying capacity of their environments. The results of this course of action are predictable and catastrophic. Once the carrying capacity of the environment has been exceeded, the death rate of the population rises in excess of the birth rate, and the vast majority of the population dies.

Figure 10.13

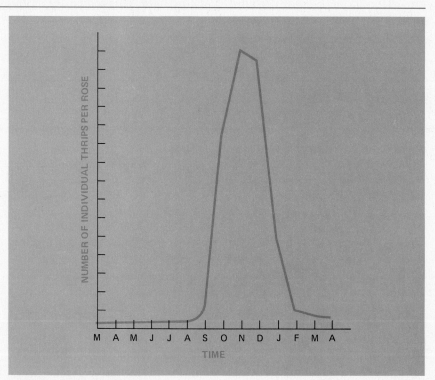

A J-shaped growth curve.

More often than not the population dies back to a level far below that of the previous carrying capacity. This situation results because the overly large population size has grossly modified the environment to the point where resources have been eliminated and the carrying capacity has been lowered drastically as a consequence.

The alternating periods of boom and bust exhibited by populations with J-shaped growth curves must correspond in some way with alternating changes in their environments. However, we do not fully understand what these changes are. That such cycles can be maintained repeatedly implies that explosive growth coupled with massive mortality can be an effective long-range means of using the environment. It is apparent that some communities in nature are largely composed of populations that exhibit J-shaped growth curves. The fact that they are not very common may be due to the inherently unstable nature of their existence.

The Ecological Status of People

The impression that I have tried to convey consistently in this chapter, and to a lesser degree in many of the previous chapters, is that living things are very intricately organized and very closely interdependent. Human beings are not excepted from this generalization. Living organisms form a very complex web of interactions, so much so that an action directed toward one small segment of an ecosystem has reverberating effects upon many other parts of the ecosystem. Often these effects may have been unintended, and perhaps totally unexpected. The classic, and perhaps most bizarre, example I can think of is a case of trading malaria for **bubonic plague.**

Several years ago the health authorities of Borneo decided to try to eliminate malaria in one region of their country where it was incapacitating large numbers of people. Knowing that malarial organisms are spread by mosquito bites, the authorities waged an all-out campaign against the mosquitoes by spreading **DDT** from airplanes. This insecticide not only poisoned and killed many mosquitoes, but it also poisoned several species of lizards. The prime effect of DDT poisoning in lizards is to promote uncoordinated movements before death. These twitchy lizards, carrying their burden of DDT, became conspicuous and easy prey for the local house cats. Consuming poisoned lizards killed the cats. Without the cats to patrol the villages, the population of rats increased. Some of the rats carried fleas that were infected with the microorganisms responsible for bubonic plague. As the rat, flea, and plague organisms increased in the villages, some of the fleas bit people and infected them with the disease. The health authorities now had a new problem with which to deal. To add insult to injury, a number of species of **caterpillars,** whose population size had previously been kept small by the predation of the lizards, were now able to increase in numbers. They devoured the materials used to thatch the roofs of village houses, resulting in leaky roofs that would not keep out the seasonal rains.

This particular story may be extreme to the point of being absurd, but it does illustrate the basic principle of interdependence of living things. It is extremely difficult to manipulate or modify living communities without producing multiple effects, many of which may be directly deleterious to the manipulator or modifier. Much of civilized humanity's activities have been carried out under the tacit assumption that it is our right and responsibility to modify ecosystems, and that we can do so without suffering any kinds of unfortunate consequences. History has repeatedly demonstrated the fallacy of this thinking, but we persist.

Many of the major human problems of today have a large biological component. Perhaps by using some of the knowledge achieved in the scientific study of life, we can obtain some insights on potential solutions to these problems. The extent to which we can live with the generalizations discussed within this book may very well affect the continuation of the biosphere. Whether or not we take these ideas seriously, our lives will be profoundly changed.

As just one example of the kind of thing that a biologist may consider as a major human problem, examine the graph in Figure 10.14. This graph is a simplified depiction of the population growth of people for roughly the last 4000 years. For approximately the last three centuries the human population has experienced unprecedented growth. This growth surge is almost entirely a result of a dramatic decrease in the death rate. The death rate has declined globally because of the influences of modern medicine and public health measures, which in large part have been brought about by our increased understanding of microorganisms. At the same time the birth rate has declined only slightly, resulting in an annual growth rate of approximately 2 percent. The world population is now expanding at a rate that will about double its size every 35 years.

Knowing the rate of growth makes it possible to go through all sorts of mathematical exercises designed to inform us about the projected size of the human population at some subsequent time. Just like the story about the mythical young man who worked for double his salary each day, the numbers become very large, very quickly. The simplest projections work on the assumption that the population growth

Figure 10.14

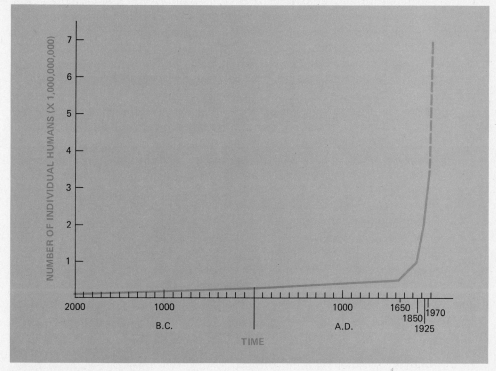

Human population growth.

rate will remain the same, and they ignore the kinds of interactions that we know occur between populations and their surroundings.

Exponentially expanding populations usually fit either one of the two patterns already described, S-shaped or J-shaped. The important difference between these two patterns is how the population growth responds to the carrying capacity of the environment. Populations with S-shaped patterns bring their death and birth rates into equilibrium before reaching the carrying capacity of the environment, thus maintaining stable population size at the environmental limits. Populations with J-shaped patterns expand beyond the carrying capacity of the environment, exceeding its limits until massive mortality reduces it to a minimum supportable size.

It would be most pertinent to inquire about the carrying capacity of the earth for people in order to determine what the graph will show in future years. It is nearly impossible to give a firm answer to such a question in light of two features of humanity. The first variable is that all humans do not live at the same standard, which has a strong effect on the amount of resources needed to maintain them. The second variable is that humans, historically, have devised all kinds of clever ways of raising the carrying capacity of their immediate environments. A further exploration of each of these variables is required.

The American standard of living is something with which each of us is familiar. The American population presently accounts for between 5 and 6 percent of the total world population. Depending upon which expert you wish to cite, Americans use somewhere between 30 and 50 percent of the world's resources. The remaining 95 percent of the population must use the remaining 50 to 70 percent of the resources.

If we make some simplistic assumptions and apply elementary mathematics, we can calculate that the carrying capacity of the earth must fall somewhere between 500 million and 8 billion people, depending upon the standard of living. What this means is that if all the people lived at the U.S. standard—or consumed resources at the rate we consume them—then the earth could probably support around 500 million people. But if everyone lived at some average non-American standard, the earth could support around 8 billion. The present world population is around 4 billion people.

If we wish to have everyone in the world live as we do today, then either we must find eight times as many resources as we now know exist, or we must eliminate $3\frac{1}{2}$ billion people who are already on the planet. Alternatively, we can probably support twice as many people as are now living if the population of the United States greatly lowers its consumption of resources. Any of these courses of action requires major changes of thinking.

Much of human effort has been devoted to finding ways of removing environmental limits—raising the carrying capacity of the planet. All the basic industries are concerned with finding or producing more food and other materials used to keep more individual people alive. There have been many notable successes in this direction, as evidenced by the rapid increase in population size over the time period that coincides with industrialization and the increased use of technological knowledge.

Our success in raising the carrying capacity of the planet for more people has been achieved at some expense. The environment we inhabit—land, water, and air—has been increasingly modified by the products of our activities. Streams become polluted; land is lost to erosion or to construction of buildings, roads, and so forth, or made unusable by mining operations; air becomes filled with molecules that are harmful to human health. Natural cycles can and do correct many of these problems with enough time, or human beings can correct them if they expend enough energy. Neither of these resources appear to be very abundant.

In fact, many resources are becoming much more difficult to obtain in sufficient quantities to meet human needs. In spite of the astounding increases that have been accomplished in food production, for example, the number of people in the world who are undernourished or malnourished grows just as steadily. There are more hungry people in the world today than there have ever been in the past.

These conditions, an increase of detrimental waste products and a decrease in available resources, both indicate that the population is approaching the environmental carrying capacity. What most populations do under these conditions is to bring birth rate and death rate into equilibrium by either lowering the birth rate or raising the death rate, or effecting some combination of the two. In this way the population size remains constant and the carrying capacity is not exceeded. Either method—increasing the death rate or decreasing the birth rate—is equally effective in changing the rate of population growth.

Most people are disinclined to seriously suggest raising the death rate, for obvious reasons. Human life is regarded very highly, and there are strong conventions against terminating it. Most people recognize that modifying these conventions would significantly lessen the chances of their own continued survival. Strong ethical arguments are also raised against most means of decreasing the birth rate. The most usual attitude is that other people should be controlling their birth rates. This leaves us at something of a behavioral standoff—and the population continues to grow.

As I have already noted, the results of exceeding the carrying capacity of the environment are profound. If we don't devise some method of preventing ourselves from exceeding this critical size, massive mortality will occur inevitably. Not finding a solution to this problem guarantees that the environment will impose its own solution. It is a certainty that the world of the future will be very different from the world of today.

Summary

1 / *Trophic levels are categories of organisms grouped together on the basis of their food sources. The most common levels or categories are producers, herbivores, carnivores, secondary carnivores, and decomposers.*

2 / *Communities may consist of simple food chains where one species feeds*

on another species at the next lower trophic level and is, in turn, fed upon by a third species in the next higher trophic level. More commonly, communities have much more complex feeding relationships, called food webs.

3 / Most communities have the greatest numbers and weights of living material at the lowest trophic level, with ever-diminishing numbers and weights in successively higher trophic levels. These observations have been described as the pyramids of numbers and biomass.

4 / All communities exhibit a pyramid of energy; the largest amount of biologically usable energy is found at the lowest trophic level and progressively less energy is available at higher levels.

5 / The pyramids of numbers, biomass, and energy all seem to be consistent with the second law of thermodynamics, which states that energy transformations are never completely efficient.

6 / Available energy flows through trophic levels (or ecosystems) in ever-diminishing amounts.

7 / The three most commonly recognized forms of interspecific interactions are predation or parasitism, competition, and mutualism.

8 / Natural selection tends to minimize the negative effects of interspecific interactions over long periods of time.

9 / Ecosystems function in such a way that the transformation and cycling of matter is a natural result. Biogeochemical cycles establish a subtle set of indirect interdependencies among almost all the species in an ecosystem.

10 / Many species of organisms have strong tendencies to change their surroundings in ways that are detrimental to themselves and beneficial to the existence of some other species. This tendency results in the orderly replacement of one community by another with a different species composition, a transformation called community succession.

11 / The dynamics of population growth are largely determined by three properties of the population: birth rate, death rate, and longevity.

12 / Rates of growth are geometric progressions, increases (or decreases) by some constant multiplier.

13 / Most populations exhibit a sigmoid (S-shaped) growth curve in which the population expands geometrically until achieving some absolute size. At that point the population size remains relatively constant.

14 / The total amount of resources available in the surroundings establishes the carrying capacity, the absolute size of a population that can be supported in that environment. The carrying capacity determines the limits of a population's growth.

15 / Some populations exhibit J-shaped growth curves in which they appear to grow beyond the carrying capacity of their environments and suffer massive mortality as a direct consequence.

16 / Ecosystems are composed of so many intricate interdependencies that it is nearly impossible to impose a simple change without producing a multitude of effects.

17 / Human populations have grown explosively within approximately two

hundred years with predictable consequences. It is obvious that a continuation of this trend cannot be maintained.

Questions to Think About

1 / Name the trophic level in which you would classify the following species: goldfish, parakeet, goldenrod, robin, red bat, Venus's flytrap, mushroom, leopard, snake-eating eagle, tuna.

2 / An exception to the usual pyramid of numbers is found in cave communities. These communities are almost entirely composed of various levels of consumers, because green plants cannot grow in total darkness. How do the consumers survive? Is your answer altered by the knowledge that some cave dwellers never leave the caves?

3 / Why is it that the economically poor, densely populated countries of the world tend to have diets containing far less meat than human diets in the United States?

4 / It is easy to see how different species of animals interact in ways that can be described as predatory or parasitic, competitive, or mutualistic. Can you give examples of ways in which plantlike organisms interact in each of these categories?

5 / What would the world be like if there were no decomposers of any kind? How long could life continue?

6 / Are human beings members of a climax community, or are they a part of some earlier successional stage?

7 / How many activities have you performed this week that could be considered as direct tinkering with the environment? How many things have you done that required indirect tinkering with the environment? For those of you who came up with fewer than five activities, think again.

Glossary

abiotic Nonliving.

acetic acid A colorless, pungent liquid, $C_2H_4O_2$, usually obtained by the destructive distillation of wood or the oxidation of alcohol. Vinegar is a dilute and impure acetic acid.

adaptation 1) The acquiring of characteristics by a species that make it better suited to live and reproduce in its environment. 2) A peculiarity of structure, physiology, or behavior of an organism that especially aids the organism in its particular environment.

adenine An alkaloid of the purine series, $C_5H_5N_5$; a nitrogenous base; formerly known as vitamin B_4. See Figure 6.7.

adenosine triphosphate A complex molecule composed of adenine, ribose, and three phosphates; the universal energy currency; also known as ATP. See Figure 9.4.

aerobic With or requiring free oxygen.

alcaptone A cyclic organic acid, $C_8H_8O_4$; an intermediate product in the metabolic breakdown of the amino acid tyrosine; it turns black on exposure to air. Also spelled *alkapton;* also known as homogentisic acid.

alcaptonuria A hereditary disease in human beings characterized by dark-colored urine.

alga (pl. algae) A photosynthetic eucaryotic organism lacking multicellular reproductive organs.

allele One of the alternate forms of the same functional gene. Alleles occupy the same position (site) on homologous chromosomes and so are separated in meiosis.

allopolyploidy Being a polyploid in which the different sets of chromosomes come from different species or widely different strains.

ambiguous Having a double meaning; capable of being understood in more than one way.

amino acid Organic compounds that contain an organic acid group and an amino group; the repeating subunit of protein molecules. See Figures 6.10 and 6.11.

amino group $-NH_2$.

ammonia A colorless, pungent, suffocating gas, NH_3, obtained chiefly by the catalytic synthesis of nitrogen and hydrogen.

ammonification The formation of ammonia, as in the soil, by the action of microorganisms upon nitrogenous organic substances; also known as nitrogen fixation.

amoeboid Moving or eating by the means of temporary cytoplasmic extensions of the cell body; like an amoeba.

anabolic Of or pertaining to the buildup of materials; constructive metabolism; opposite of *catabolic.*

anaerobic Without free oxygen.

analogy 1) Agreement or resemblance in certain aspects, as form or function, between otherwise dissimilar things; similarity without identity. 2) Reasoning in which relations or resemblances are inferred from others that are known or observed; reasoning that proceeds from the individual or particular to a coordinate individual or particular. 3) A similarity in function or superficial appearance but not in origin.

androgens Any of various hormones that control the appearance and development of masculine characteristics.

antecedent (*adj.*) Going or being before, preceding; (*n.*) one who or that which precedes or goes before.

antheridia (**sing. antheridium**) The male reproductive organs in simple plants.

anticodon Three adjacent nucleotides in a tRNA molecule that are complementary to the three nucleotides of a codon in mRNA.

archegonia (**sing. archegonium**) Multicellular egg-producing organs in simple plants.

asexual Without sex.

atom The smallest unit a chemical element can be divided into and still retain its characteristic properties.

autopolyploidy Being a polyploid where all the chromosomes are from the same source.

autosome Any chromosome except a sex chromosome. Human beings have 22 pairs of autosomes.

autotrophic Self-nourishing; said especially of plants that make their own food by photosynthesis and of bacteria capable of growing without organic carbon or nitrogen.

axillary Located within the angle formed between a leaf and the stem.

bacteriophage A virus that can destroy bacteria and induce hereditary changes.

bacterium (pl. bacteria) A small unicellular organism that lacks a formed nucleus (the genetic material may be dispersed throughout the cytoplasm).

basal temperature The minimum temperature maintained while the body is at complete rest.

binary fission A form of asexual reproduction found in procaryotic cells.

binomial expansion Increasing an algebraic expression consisting of two functions by some exponential or power; for example, $(a + b)^2 = a^2 + 2ab + b^2$

binomial nomenclature A formal name for a species consisting of two latinized words.

biogeochemical cycles The repeated interconversion of different kinds of matter brought about by living and nonliving factors.

biomass The weight of the living material in an organism or a community.

biosphere The global ecosystem; the shell of living material surrounding the planet.

biotic Pertaining to any aspect of life; living.

birth rate The number of new individuals added to the population per some standard number of individuals already in the population within some standard unit of time. For example, 15 percent per year = 15 births per 100 individuals per year.

blastocyst An early stage in the development of mammals, achieved at about the time the fertilized egg reaches the uterus and burrows into the uterine lining.

bond That which binds or holds together; a unit of combining power between the atoms of a molecule; a connection between two atoms to form a molecule. A high-energy bond is a connection between two atoms that when broken releases more energy than other normal bonds.

botanist A person who studies plants.

breathing A process that moves large volumes of gases in and out of specialized organs such as lungs or gills so that gas exchange may occur.

bubonic plague A malignant, contagious, epidemic disease characterized by fever, vomiting, diarrhea, and inflamed swelling of lymph glands in the groin or armpit. It is caused by a bacterium transmitted to humans by fleas from infected rats.

Calorie The amount of heat energy required to raise the temperature of one kilogram of water by one degree Celsius (1°C).

carbohydrate Organic compounds with the general formula $(CH_2O)_n$, including sugars, starches, and cellulose.

carnivore A meat-eating organism, or a secondary consumer.

carrying capacity The number of organisms whose resource requirements can be met in a particular environment; a measure of the environment's ability to support life.

cartilage A tough elastic form of connective tissue composed of cells embedded in a translucent matrix, either homogeneous or fibrous; gristle.

catabolic Of or pertaining to the breaking down of materials; destructive metabolism; opposite of *anabolic*.

catalyst A substance that accelerates a chemical reaction but is not used up in the reaction. Enzymes are catalysts.

caterpillar Larval form of butterflies and moths.

cell The fundamental unit of structure and function of all living things, consisting of a small, usually microscopic mass of cytoplasm, variously differentiated, and surrounded by a semipermeable membrane.

centi- Prefix meaning "one-hundredth"; $\frac{1}{100}$; $\times 10^{-2}$.

centromere The point on a chromosome, usually seen as a constriction, to which the spindle fibers attach during replication. It is not necessarily located in the center of the chromosome.

chiasma (pl. chiasmata) An X-shaped connection between paired homologous chromosomes at meiosis. See Figure 5.13. Also called a *crossover*.

chlorophyll The green nitrogenous coloring matter contained in the chloroplasts of plants, essential to the production of carbohydrates by photosynthesis. It occurs in two forms: the bluish-green chlorophyll A, $C_{55}H_{72}O_5N_4Mg$, the most common form; and the yellowish-green chlorophyll B, $C_{55}H_{70}O_6N_4Mg$.

chorionic gonadotropin A hormone produced by membranes of mammalian embryos that mimics the effects of LH in that it stimulates the production of estrogens and progesterone.

chromatid Each of a pair of new sister chromosomes from the time at which molecular duplication occurs until the time at which the centromeres separate during anaphase.

chromatin The nucleic acid-protein complex found in eucaryotic chromosomes.

chromosome A complex structure found in the nucleus of a eucaryotic cell, composed of nucleic acid and proteins and bearing the genetic information of the cell.

classify To arrange or put into groups or classes by certain resemblances.

climax A self-perpetuating community of stable species composition.

cloning Deriving a group of cells or individuals from a single individual through repeated asexual reproduction.

coacervate A mixture of chemical substances that segregates into two phases: a droplet phase, which is randomly dispersed but highly organized internally; and a continuous matrix phase, which is not organized in any way.

codon Three adjacent nucleotides on a molecule of DNA or mRNA that form the code for a single amino acid.

coenzyme A A chemical compound derived from pantothenic acid that helps to accomplish the movement of pyruvate molecules into the Krebs cycle.

colloid A state of matter in which finely divided particles of one substance are suspended in another in such a manner that the electrical and surface properties acquire special importance.

colony A group of cells, usually derived from the

same parent cell, functioning in close association but with only limited dependence on one another.

comb   The fleshy crest on the head of domestic fowl.

community   A group of species living together in a definite area.

competition   An interaction wherein two individuals or species attempt to use the same resource, which is in limited supply.

compound   Any substance composed of two or more different elements bonded together.

concept   A mental image, especially a generalized idea, formed by combining the elements of a class into the notion of one object.

condom   A sheath for the penis, usually made of rubber and having an antivenereal or contraceptive function.

consumer   Any organism that eats or consumes another organism for food; a heterotroph.

control   A standard of comparison against which to check the results of a scientific experiment; also called *control group.*

corpus luteum   The hormone-secreting body into which an ovarian follicle is converted immediately after ovulation; literally means "yellow body."

crossing-over   The mechanism by which linked genes undergo recombination. In general the term refers to the reciprocal exchange of corresponding segments between two homologous chromatids.

crossover   An X-shaped connection between paired homologous chromosomes during meiosis. See Figure 5.13.

cryptic   Concealing; protecting.

crystal   A homogeneous solid body exhibiting a definite and symmetrical internal structure, with geometrically arranged cleavage planes and external faces that assume any of a group of patterns associated with peculiarities of atomic structure.

cytochrome   Iron-containing red proteins; components of the electron transport machinery in respiration.

cytokinesis   The division of cytoplasm in a replicating cell.

cytoplasm   The contents of a cell, usually excluding the nucleus.

cytosine   An alkaloid of the pyrimidine series, $C_4H_5N_3O$; a nitrogenous base. See Figure 6.6.

DDT   A powerful insecticide effective on contact: dichlorodiphenyltrichloroethane.

death rate   The number of individuals lost from a population by death per some standard number of individuals already in the same population within some standard unit of time. For example, 15 percent per year = 15 deaths per 100 individuals per year.

deca-   Prefix meaning "ten"

deci-   Prefix meaning "one-tenth"; $\frac{1}{10}$; $\times 10^{-1}$.

decomposer   An organism of decay or putrefaction; one that breaks down organic molecules usually external to itself.

denitrifying   Reducing nitrates to nitrites, nitrogen, or ammonia.

development   The innate changes in structure and function that occur during the life of an organism.

diaphragm   1) A muscular wall that separates the thoracic and abdominal cavities of mammals. 2) Any membrane or partition that separates or divides. 3) Any device resembling a diaphragm in appearance or elasticity. 4) A contraceptive device of soft rubber placed at the junction of the vagina and uterus.

differentiate   To cause to become unlike; develop differences.

differentiation   The process of developing differences.

diffusion   The spreading out in all directions of any fluid, caused by thermal agitation.

dihybrid cross   Mating of two individuals that have differing phenotypes of two features. A cross involving two different genes.

dipeptide   Two amino acids joined by a peptide bond.

diploid   Having two sets of chromosomes; having two of each kind of chromosome.

disjunct   Separate, detached.

DNA   A polymeric macromolecule that is the genetic material; deoxyribonucleic acid. See Figures 6.5, 6.7, and 6.8.

dominant   Designating one of a pair of hereditary

expressions of character that, appearing in heterozygous offspring, masks an alternative expression.

ecology The scientific study of the interaction of organisms with their environment, including both the physical environment and the other organisms that live in it.

ecosystem The organisms of a particular habitat, such as a pond or forest, together with the physical conditions in which they live.

egg A female gamete in animals; ovum.

electron One of the three most fundamental particles of matter, with almost no mass and an electrical charge of -1.

electron transport The movement of high-energy electrons from one molecule to another with a consequent loss of energy. In most living systems this movement is accomplished by cytochromes, the derived energy being used to add phosphates to ADP molecules, producing ATP molecules.

element One of a limited number of substances, each of which is composed entirely of atoms having an invariant nuclear charge, and none of which may be decomposed by ordinary chemical means.

embryo The earliest stages in the development of an organism, before it has assumed its distinctive form.

emigration The movement of individuals out of a population.

endocrine gland One of several ductless glands, such as the thyroid, whose secretions, released directly into the blood or lymph, have a critical importance in many phases of normal living activity.

endoplasmic reticulum A system of folded membranes, often with ribosomes attached, found in the cytoplasm of eucaryotic cells.

energetics The science of the laws and phenomena of energy.

energy The capacity for doing work and for overcoming inertia.

enucleated When the nucleus has been removed.

enzyme A globular protein that functions as a catalyst.

epithelial cells Those cells that line the canals, cavities, and ducts of the body or the free surfaces exposed to air.

equator A line lying in a plane perpendicular to a sphere's polar axis.

estrogen Any of various substances that influence or produce changes in the sexual characteristics of female animals; also spelled *oestrogen*.

estrous The peak of the female sexual cycle, culminating in ovulation, found in most mammalian species; in heat; in season; the fertile period; also spelled *estrus*.

estrous cycle The entire cycle of physiological changes in most species of female mammals, preparing the reproductive organs for their fertile period.

ethanol An alcohol, C_2H_5OH, obtained by the distillation of certain fermented sugars or starches; the intoxicant in liquors, wines, and beers; also called *ethyl alcohol* or grain alcohol.

eucaryotic Having the genetic material inside a nucleus; also spelled *eukaryotic*.

evolution Any gradual change. Organic evolution, often referred to as evolution for short, is any genetic change incorporated into populations from generation to generation.

excretion The elimination of nitrogenous waste matter by normal discharge from an organism or any of its tissues.

experiment (*n.*) An act or operation designed to discover, test, or illustrate a truth, principle, or effect; (*v.*) to make a test or trial.

experimentation The act or practice of experimenting.

experimental group A population whose treatment differs from that of the control group by only one factor.

exponential growth Growth, especially in the number of organisms in a population, that is a simple function of the size of the growing entity: the larger the entity, the faster it grows.

extrinsic Not included or essential; being outside the nature of something; opposite of *intrinsic*.

FAD A complex organic compound, derived from riboflavin, that functions as part of the electron transport system in aerobic respiration; *f*lavin *a*denine *d*inucleotide. See Figure 9.11.

fallopian tube   One of a pair of long slender ducts serving as a passage for the ovum from the ovary to the uterus; also called *oviduct*.

fermentation   Generally, the degradation of a molecule such as glucose to smaller molecules with the extraction of energy without the use of oxygen; more precisely, the reaction of pyruvate with reduced NAD to form either CO_2 and ethanol or lactate and to release nonreduced NAD.

fertilization   The fusion of male and female gametes; impregnation.

Feulgen stain   A biological stain that reacts specifically with nucleic acid to produce a distinctive red color.

filial   Pertaining to a generation following the parental.

***flagella* (sing. *flagellum*)**   Long whiplike appendages used to propel procaryotic or eucaryotic cells.

flagellated   Having flagella.

follicle   A small cavity or sac in certain parts of the body having a protective or secretory function; for example: hair follicle, ovarian follicle.

follicle-stimulating hormone   Pituitary hormone that in females causes an ovarian follicle to increase in size and in males promotes gametogenesis; usually called FSH.

food chain   A simple sequence of prey species and the predators that consume them.

food web   The complete set of food links between species in a community; a diagram indicating which species are the eaters and which are the consumed.

frequency   1) The number of times something occurs in relation to the total number of possible occurrences. 2) The number of times a given case, value, or event occurs in relation to the total number of classified cases, values, or events.

Fungi   A kingdom of organisms that includes any of a number of nonflowering, nonphotosynthetic, plantlike organisms; usually reproduce asexually and often grow parasitically or on dead organic matter; examples include mushrooms, molds, and mildews.

gamete   A mature, haploid, sexual reproductive cell; an egg or sperm.

gametogenesis   The specialized series of cellular divisions that lead to the production of sex cells; meiosis.

gas exchange   The reciprocal diffusion of two gases (usually oxygen and carbon dioxide) across membranes, as in a lung.

gene   A unit of heredity; often used as the unit of genetic function that carries the information for a single polypeptide.

gene pool   All the alleles of all the genes in a population.

genetic   Hereditary.

genetics   The science dealing with the interaction of the factors in producing similarities and differences among individuals related by descent.

genetic equilibrium   A condition where allelic frequencies remain constant generation after generation.

genotype   A symbolic description of the genetic constitution of an individual with respect either to a single trait or to a larger set of traits.

***genus* (pl. *genera*)**   A group of closely related species.

geometric progression   A sequence of terms of which each member is greater than its predecessor by a constant ratio, as 2, 4, 8, 16, 32, 64.

glucose   A simple sugar containing six carbon atoms. See Figure 9.5.

glycolysis   The enzymatic breakdown of glucose to pyruvate under anaerobic conditions.

Golgi apparatus   A system of concentrically folded membranes found in the cytoplasm of eucaryotic cells that plays a role in the production and release of secretory materials such as digestive enzymes.

gonad   An organ that produces gametes.

gonadotropin   Any hormone that stimulates the gonads, such as FSH or LH.

gonadotropin-releasing substance   Hypothalamic substance that stimulates the pituitary to produce and release gonadotropins.

growth   A gradual increase in size.

guanine   An alkaloid of the purine series, $C_5H_5N_5O$; a nitrogenous base. See Figure 6.6.

haploid   Having a chromosome complement consisting of just one copy of each kind of chromosome.

hecto- Prefix meaning "one hundred"; $\times 10^2$.

helical Shaped like a helix.

helix A line, thread, wire, or the like, curved into a shape such as it would assume if wound in a single layer around a cylinder; a form like a screw thread or a Slinky.

heme The insoluble, protein-free, iron-containing constituent of hemoglobin, $C_{32}H_{32}FeN_4O_4$.

hemoglobin The iron-containing protein in the blood that carries oxygen.

hemophilia A disorder characterized by uncontrolled bleeding even from slight injuries, typically affecting males; caused by a malfunction of the blood-clotting mechanism.

herbivore A plant-eating organism; a primary consumer; a vegetarian.

hereditary Transmitted or transmissible directly from a living organism to its offspring.

heredity Transmission of genetically determined characteristics from parents to offspring.

heterotrophic Requiring preformed organic materials as food.

heterozygous Carrying two different alleles of the same gene.

hierarchical Of, belonging to, or characteristic of a series of groupings in graded order.

homogentisic acid See *alcaptone*

homologous 1) Similar or related in structure, position, proportion, value, and so on; corresponding in nature or relationship. 2) Corresponding in structure or origin, as an organ or part of one organism to a similar organ or part of another.

homozygous Having identical alleles of a given gene on both homologous chromosomes.

hormone An internal secretion produced in and by one of the endocrine glands and carried by the bloodstream or body fluids to other parts of the body where it has a specific physiological effect.

hybrid Anything of mixed origin or of incongruous or different elements.

hybrid offspring The progeny of two different-appearing parents.

hypothalamus A group of structures forming a part of the middle brain, controlling visceral activities, regulating body temperature and many metabolic processes, and influencing certain emotional states.

hypothesis A scientific conclusion with limited support drawn from known facts and used as a basis for further investigation or experimentation.

immigration The movement of individuals into a population from elsewhere.

independent assortment A segregation of alleles of genes into gametes that is not affected by the simultaneous segregation of alleles of other genes.

inter- Prefix meaning "between."

interstitial cells Those cells situated between the seminiferous tubules that produce androgens.

intra- Prefix meaning "within."

intracellular Within cells.

intrinsic Belonging to or arising from the true or fundamental nature of a thing; essential; inherent.

invagination The process of infolding so as to form a depression or pouch.

isotopes Atoms with the same number of protons and electrons but differing numbers of neutrons. Thus the atoms are of the same element but have different weights. Radioactive isotopes are those that emit radiations detectable by various kinds of devices, such as Geiger counters or scintillation counters.

kilo- Prefix meaning "thousand"; $\times 10^3$.

Krebs citric acid cycle A circular chemical pathway whereby a derivative of pyruvate is broken down into carbon dioxide and water with a release of large amounts of energy; also called the **citric acid cycle.** See Figure 9.13.

lactate A salt or ester of lactic acid.

lactation 1) The production of milk. 2) The act of suckling young.

lactic acid A limpid, syrupy acid, $C_3H_6O_3$, with a bitter taste; present in sour milk.

law A formal statement of the manner or order in which a set of natural phenomena occur under certain conditions.

lens A piece of glass or other transparent substance, bounded by two curved surfaces, or by one curved and one plane surface, by which rays of light are made to converge or diverge.

Liebig's law   Essential resources present in amounts most closely approximating the amounts needed will act as the limit to the growth of a population.

ligament   A band of firm, fibrous tissue forming a connection between bones or serving to support an organ.

lineage   Those descended from the same ancestor; ancestry; pedigree; family; stock.

linked genes   Those genes that are located on the same chromosome.

lipid   Any of a large class of organic compounds, insoluble in water and typically greasy to the touch, including the fats, waxes, and steroids.

longevity   The tendency to live long.

lung   A saclike organ of gas exchange located in the chest cavity of humans and other air-breathing vertebrates.

luteinizing hormone   A pituitary gonadotropic hormone that promotes the secretion of male and female sex hormones. Usually called LH; also known as interstitial-cell stimulating hormone (ICSH).

luteotropic hormone   A pituitary hormone that maintains the activity of the corpus luteum and stimulates lactation; also called LTH or *prolactin.*

macromolecule   A giant molecule, such as a protein, rubber, cellulose, or nucleic acid.

malaria   Any of several forms of a disease caused by certain species of parasitic protozoans introduced into the blood by the bite of an infected mosquito. Its symptoms include chills, fever, and profuse sweating.

map unit   An arbitrary measure of distance between genes on the same chromosome obtained from the frequency of their recombination.

matter   Anything that occupies space and has mass.

mega-   Prefix meaning "million"; $\times 10^6$.

meiosis   Cell division of a diploid cell to produce four haploid cells. The process consists of two successive cell divisions with only one cycle of chromosome replication.

membrane   A thin, pliable, sheetlike cover or lining.

menstrual cycle   The entire cycle of physiological changes in females of most species of primates, which prepares the reproductive organs for their fertile period. It is characterized by menstruation.

menstruation   The periodical flow of bloody fluid from the uterus; occurs at roughly monthly intervals and does not coincide with ovulation.

messenger RNA   The RNA that carries genetic information from the gene to the ribosome, where it determines the order of the amino acids in the formation of a polypeptide.

metabolism   The aggregate of all chemical processes constantly taking place in a living organism.

metaphysical   1) Of or pertaining to ultimate reality or basic knowledge. 2) Beyond or above the physical or the laws of nature; transcendental. 3) Pertaining to all speculative philosophy.

micro-   Prefix meaning "one-millionth"; 1/1,000,000; $\times 10^{-6}$; also a prefix used to denote something small.

microorganism   Any organism such as a bacterium that requires the use of a microscope to be seen.

microscope   An optical instrument consisting of one or more lenses, used for enlarging the appearance of objects too small to be seen or clearly observed by ordinary vision.

mildew   A disease of plants usually caused by a parasitic fungus that deposits a whitish or discolored coating.

milli-   Prefix meaning "one-thousandth"; $\frac{1}{1000}$; $\times 10^{-3}$.

mitochondrion (pl. *mitochondria*)   An organelle that occurs in eucaryotic cells and contains the enzymes of the Krebs cycle and the molecules of the electron transport system.

mitosis   Cell division in eucaryotes leading to the formation of two daughter cells, each with a chromosome complement identical to that of the original cell.

mold   Any of a variety of fungus growths commonly found on the surfaces of decaying food or in warm, moist places.

molecule The smallest part of an element or compound that can exist separately without losing the physical or chemical properties of the original element or compound.

Monera A kingdom of organisms characterized by having procaryotic cells; the bacteria and blue-green algae.

monohybrid cross Mating between individuals differing by only one character. More particularly, it is a cross between individuals each homozygous for a different allele of the same gene.

multicellular Consisting of, or composed of, more than one cell.

mutant An organism differing from its parents in one or more characteristics that are heritable.

mutation Any discontinuous change in the genetic constitution of an organism; any heritable change.

mutualistic Having advantages to both parties.

mutualism Any interrelationship, such as that exhibited by fungi and algae in forming lichens, in which both species benefit from the association.

NAD A very complicated organic compound that has a strong attraction for hydrogen atoms, which, under aerobic conditions, may then be passed into an electron transport system; *n*icotinamide-*a*denine *d*inucleotide; also known as *d*iphospho*p*yridine *n*ucleotide or DPN.

nano- Prefix meaning "one-billionth"; 1/1,000,000,-000; $\times 10^{-9}$.

natural selection The differential contribution of offspring to the next generation by various genetic types belonging to the same population; the mechanism of evolution proposed by Charles Darwin.

negative feedback A return of information to or near the onset of a chain of reactions resulting in a reduction in the reactions of the chain.

neutron One of the three most fundamental particles of matter, with a mass of one Dalton and no electrical charge.

niacin A colorless, water-soluble compound, $C_6H_5NO_2$, prepared by the oxidation of nicotine and forming part of the vitamin B complex; used to prevent pellagra; also called *nicotinic acid.*

nitrate A salt or ester of nitric acid; usually represented as $-NO_3$.

nitrification The oxidation of ammonium salts into nitrites and nitrates, especially by soil bacteria.

nitrite A salt of nitrous acid; usually written as $-NO_2$.

nitrogenous base A nitrogen-containing compound that can neutralize an acid to produce a salt; a nitrogen-containing compound that yields hydroxyl ions $(-OH^-)$ in solution.

nuclear membrane The envelope, consisting of two layers of unit membrane, that encloses the nucleus of eucaryotic cells.

nucleic acid A long-chain, alternating polymer of pentose and phosphate groups, with nitrogenous bases as side chains.

nucleolus A small clear body, generally spherical, found within the nucleus of eucaryotic cells. It is the site of synthesis of ribosomal RNA.

nucleotide The basic chemical unit in a nucleic acid consisting of one of four nitrogenous bases linked to a pentose, which, in turn, is linked to a phosphate group.

nucleus 1) The dense central portion of an atom, made up of protons and neutrons, with a positive charge. 2) The centrally located chamber of eucaryotic cells that is bounded by a double membrane and contains the chromosomes; the information center of the cell.

optimal Producing or conducive to the best results.

organ A formed body, such as the heart, liver, brain, root, or leaf, composed of different tissues, integrated to perform a distinct function for the body as a whole.

organelle A small, often membrane-bound structure found inside of a cell.

organic acid group A functional group of atoms that usually acts as an acid in solution; usually written as COOH, or $\text{C}{\overset{\displaystyle O}{\underset{\displaystyle OH}{\Big\backslash}}}$

organic compounds Compounds containing carbon.

organism Any living thing as found in nature.

organ system A connected series of organs that together perform some unitary function, such as the reproductive system, nervous system, or circulatory system.

ovary Any female organ that produces an egg.

oviduct A Fallopian tube.

ovulate To produce an egg or ovum; to discharge an egg from an ovary.

ovulation The formation and discharge of ova (eggs).

ovule In plants an organ that contains haploid tissue and, within the haploid tissue, an egg. When it matures, an ovule becomes a seed.

ovum (pl. ova) The egg; the female reproductive cell.

oxytocin A hormone of the posterior lobe of the pituitary (probably produced in the hypothalamus) that stimulates the uterine contractions of childbirth.

pantothenic acid An unstable, oily compound, $C_9H_{17}NO_5$, widely distributed in plant and animal tissues; formerly called vitamin B_3.

parasitism The form of interrelationship in which one species lives at the expense of the other but not ordinarily to the point of killing its host.

parental phenotypes External appearances that are just like those of one or the other of the parents.

penis The copulatory organ of male animals.

pentose A sugar containing five carbon atoms.

peptide bond Connection of two amino acids caused by the reaction of the amino group of one with the acid group of the other with a consequent release of one water molecule. See Figure 6.12.

phage See *bacteriophage*.

phenotype The observable properties of an individual as it has developed under the combined influences of the genetic constitution of the individual and the affects of environmental factors.

phosphate A salt or ester of phosphoric acid. See Figure 6.4.

photosynthesis The process by which visible light is trapped and the energy used to synthesize energy-rich compounds such as ATP and glucose.

phylogeny Evolutionary relationships among organisms; the developmental history of a group of organisms.

pico- Prefix meaning "one-trillionth"; $\times 10^{-12}$.

pituitary A small rounded body, at the base of the brain in vertebrates, that secretes hormones having a wide range of effects upon the growth, metabolism, and other functions of the body.

placenta The vascular, spongy organ of interlocking fetal and uterine membranes by which the fetus is nourished in the uterus.

plastid Any of various small specialized organelles in the cytoplasm of a cell.

polar body A nonfunctional nucleus produced by meiosis accompanied by very little cytoplasm.

poles The two extremities of the axis of a sphere or any spheroidal body.

pollen The fertilizing element of flowering plants containing the male gamete.

polymer Any compound composed of simpler, repeating subunits.

polypeptide A compound composed of many amino acids joined together by peptide bonds.

polyploid Cells with many of each kind of chromosome.

population Any group of organisms capable of interbreeding, for the most part, and coexisting at the same time in the same area.

postmortem After death.

predation The act of capturing and killing other organisms for food.

primate Any mammal in the order Primates, which includes tarsiers, lemurs, marmosets, monkeys, apes, and human beings.

probability The ratio of the chances favoring an event to the total number of chances for and against it; likelihood.

procaryotic An organism without an organized nucleus in its cell or cells; also spelled *prokaryotic*.

producer An autotrophic organism; one capable of making its own food from inorganic starting materials and using an external energy source.

progesterone Hormone produced by the corpus luteum that maintains the spongy lining of the uterus.

prolactin A pituitary hormone that stimulates lactation in mammals; see also *LTH*.

prophylactic (*adj.*) Tending to protect against or ward off, as a disease; preventative. (*n.*) 1) A medicine or appliance that protects; 2) a condom.

proportion Relative magnitude, number, or degree, as existing between parts or different things.

protein Relatively large biological polymer formed from amino acids.

Protista A kingdom of single-celled or multicellular organisms with single-celled reproductive structures. This kingdom includes the protozoans, most algae, and some moldlike species.

proton A positively charged particle in the atomic nucleus, equal in mass to a neutron.

protonema The hairlike growth form that constitutes an early stage in the development of the moss.

protoorganism An arbitrary word applying to the earliest forms of life.

pubic bones Two bones that join to form the lower margin of the hip girdle and support much of the abdomen.

purine A double-ring, nitrogen-containing base that is a component of nucleic acids and several other biologically active substances.

pyramid of biomass A graphic representation of the distribution of biomass in a community by trophic level.

pyramid of energy A graphic representation of the distribution of available energy in a community by trophic level.

pyramid of numbers A graphic representation of the distribution of individual organisms in a community by trophic level.

pyrimidine A single-ring, nitrogen-containing base that is a component of nucleic acids.

random Done or chosen without definite aim or deliberate purpose; chance; casual.

reaction The reciprocal action of substances subjected to chemical change, or some distinctive result of such action.

recessive An alternative allele or character whose expression is masked by the presence of the other; opposite of *dominant*.

reciprocal cross A pair of matings in which the phenotypes of the two sexes are reversed, as, for example, red male × white female and white male × red female.

recombinant phenotypes External appearances that are a mixture of both parents, showing one trait from one parent and another trait from the second parent.

recombination The shuffling of phenotypes caused by sexual reproduction.

reducing agent A chemical substance that has a strong tendency to give up hydrogens or remove oxygen from another molecule; also known as a reducer.

reduction The process of depriving a compound of oxygen; the process by which atoms gain electrons or cease to share them with a more electronegative element.

redundant Excessive; being more than required.

relativity, special theory of The velocity of light is the maximum velocity possible in the universe; it is constant and independent of the motion of its source. Motion itself is a meaningless concept except as between two physical systems or material bodies moving relatively to each other; and energy and mass are interconvertible in accordance with the equation $E \times mc^2$, or energy equals mass times the square of the speed of light.

reproduction The process by which an organism gives rise to another of its kind.

resolution The separation of anything into component parts; the ability to distinguish the parts of a structure.

respiration The oxidation of the end products of glycolysis to produce much more usable energy than the simple glycolytic process can liberate.

riboflavin A member of the vitamin B complex (vitamin B_2); an orange-yellow crystalline compound, $C_{17}H_{22}N_4O_9PNa \cdot 2H_2O$, found in milk, leafy vegetables, egg yolk, and meats; also made synthetically; formerly called vitamin G.

ribosome A small cellular structure that is the site of protein synthesis.

rhizoid Hairlike extensions of cells in mosses, liverworts, and a few vascular plants that serve the same functions as roots and root hairs in higher plants.

RNA A nucleic acid whose nucleotide uses the pyrimidine uracil instead of thymine, and the pentose found is ribose; *ribo*nucleic *a*cid.

scrotum The pouch of skin that contains the testes of most male mammals.

secretion 1) The process, generally a glandular function, by which materials are separated from the body fluid and elaborated into new substances, such as milk, bile, or hormones. 2) Any such substances produced by the process of secretion.

seed The ovule from which a plant may be reproduced.

self-pollinating When a plant is able to fertilize itself, using its own pollen.

seminal fluid A secretion, produced by glands lining the male reproductive tract, that is used to transport and nourish the male gametes.

seminiferous tubules Cylindrical structures within the testes composed of cells that undergo meiosis to produce male gametes.

semipermeable Partially allowing passage, most particularly with respect to fluids.

sex Either of two divisions, male or female, by which organisms are distinguished with reference to the reproductive functions.

sex chromosome A chromosome whose presence in the cells of certain plants and animals is associated with the determination of maleness or femaleness. In mammals each diploid female cell contains two X chromosomes and each diploid male cell contains one X chromosome and one Y chromosome. X and Y are considered sex chromosomes.

sex-linked genes Genes that are associated with or carried on sex chromosomes. For practical reasons we have limited knowledge about genes carried on Y chromosomes of mammals.

sexual reproduction The fusion of two haploid gametes to produce one diploid zygote.

Shelford's law Most organisms have a range of tolerance for conditions and substances needed to support life. Population growth is limited by these tolerance ranges.

sigmoid S-shaped; like an S.

solution Mixture consisting of molecules or ions less than one nanometer in diameter, permanently suspended in a fluid medium (water in most biological systems).

speciation The formation of a species by the action of evolutionary processes upon a population of organisms.

species Taxonomic category consisting of a group of actually or potentially interbreeding natural populations that ordinarily do not interbreed with other such groups, even when there is opportunity to do so.

sperm Male reproductive cells, or gametes.

spermicidal Lethal to male gametes.

spindle 1) The slender rod on a spinning wheel, containing a spool or bobbin on which the thread is twisted and wound. 2) Any narrow, tapering object resembling a spindle. 3) A structure of elongated, colorless fibers formed during cellular replication.

spontaneous generation The production of life from nonliving material without the agency of already existing life.

spore An asexual reproductive cell capable of growing into an adult without fusion with another cell.

stalactite A conical incrustation hanging from the ceiling of a cavern.

stalagmite A conical incrustation on the floor of a cavern.

staminate Having pollen-bearing organs.

sterilization The act or process of removing reproductive powers, especially by surgical operation; making barren.

stimulation The act or process of excitation that influences the activity of an organism as a whole or in any of its parts.

succession (community) A sequence or series of

communities that regularly replace one another; largely caused by environmental changes produced by preceding communities.

sugar Any of a large class of carbohydrates, structurally similar to the sweet, crystalline substance $C_{12}H_{22}O_{11}$ obtained from the juice of various plants.

suspension Mixture containing solid particles larger than 100 micrometers distributed throughout a fluid; the particles will ultimately settle out under the force of gravity.

symbiotic The state of living together in close association with mutual benefit.

sympatric Occurring in the same geographic area.

synapsis The union, side by side, of homologous chromosomes early in meiosis.

synthesis The formation of a compound from other, usually simpler substances.

telescope An optical instrument for enlarging the image of distant objects.

terminator codon A sequence of three adjacent nucleotides in an mRNA molecule that signals the end of polypeptide synthesis.

test cross Mating an individual with a dominant phenotype to one that is homozygous recessive to determine the most probable genotype of the dominant phenotype.

testes (**sing. *testis***) Male gonads; male reproductive organs.

testosterone A male sex hormone, $C_{19}H_{28}O_2$, produced by interstitial cells of the testes. See Figure 8.5.

tetrad Paired homologous chromosomes during the first prophase and metaphase of meiosis. Each chromosome at these stages will be a double structure, consisting of two chromatids joined at a not-yet-divided centromere. See *synapsis*.

tetraploid A cell with four of each kind of chromosome.

theory A closely reasoned set of propositions, derived from and supported by established evidence and intended to serve as an explanation for a group of phenomena.

thermodynamics The study of the relationship between heat and other forms of energy.

thermodynamics, laws of 1) Energy can be changed from one form to another, but it cannot be created or destroyed. 2) All natural processes tend to proceed in such a direction that the disorder or randomness of the system increases.

thymine An alkaloid of the pyrimidine series, $C_5H_6N_2O_2$; a nitrogenous base. See Figure 6.6.

tissue A group of similar cells organized into a functional unit and usually integrated with other tissues to form part of an organ.

toad A tailless, hopping, insect-eating amphibian, resembling a frog but usually more terrestrial in its habits, often with warty growths on the skin.

transfer RNA The type of RNA that becomes attached to an amino acid and guides it to the correct position on the ribosome-mRNA complex for protein synthesis.

transformation 1) A change in form or appearance. 2) The chemically induced acquisition of new genetic traits.

trihybrid cross Mating between two individuals that have differing phenotypes for three different characters. More precisely, it is the cross of two individuals homozygous for different alleles at each of three different genes.

tripeptide Three amino acids joined by peptide bonds.

trophic levels Categories of species based on the most usual source of food energy.

true-breeding Of pure strain or pedigree; homozygous for all relevant genes.

tubal ligation A form of sterilization performed on females in which the fallopian tubes are tied off and cut, thus preventing the egg from reaching the uterus or sperm from reaching the egg.

unicellular Consisting of a single cell.

uracil An alkaloid of the pyrimidine series, $C_4H_4N_2O_2$, chemically analogous to thymine; a nitrogenous base found in RNA.

uterus The organ of a female mammal in which the young are protected and developed before birth; the womb.

vacuole A liquid-filled cavity in a cell, enclosed within a unit membrane.

vagina   The canal leading from the external reproductive opening in female mammals to the uterus.

vasectomy   A form of sterilization performed on males in which a portion of the tubes carrying sperm from the testes are tied off and removed.

venereal disease   Any illness transmitted by sexual contact, such as syphilis and gonorrhea.

virulent   1) Exceedingly noxious or harmful. 2) Having the power to injure an organism by invasion of tissue and generation of internal toxins.

virus   1) Any of a class of filterable, submicroscopic, disease-causing agents, composed of proteins and nucleic acid but often reducible to crystalline form; typically inert except when in contact with certain living cells. 2) Any virulent substance developed within an animal or plant body and capable of transmitting a specific disease.

viscous   Glutinous; semifluid; sticky.

vitamin   Any of a group of complex organic substances found in minute quantities in most natural foods and closely associated with the maintenance of normal physiological functions in humans and animals.

work   A transference of energy from one body to another resulting in the motion or displacement of the body acted upon, expressed as the product of the force and the amount of displacement in the line of its action.

zoologist   A person who studies animals.

zygote   A cell formed by the union of two gametes.

Index

82 83 84 9 8 7 6 5 4 3 2